Sintering of Ceramics

Sintering of Ceramics

Edited by **Carl Burt**

New York

Published by NY Research Press,
23 West, 55th Street, Suite 816,
New York, NY 10019, USA
www.nyresearchpress.com

Sintering of Ceramics
Edited by Carl Burt

International Standard Book Number: 978-1-63238-415-7 (Hardback)

Contents

Permissions

List of Contributors

Preface

The process of sintering involves forming solid mass of material by applying heat/pressure without reaching the point of liquefaction. This book covers latest techniques and methods introduced in the field of sintering with special focus on the applications of sintering for formation of ceramics. It is a compilation of the research works conducted by reputed experts in this field. Major topics related to sintering of ceramics such as conventional sintering, superconducting ceramics, dielectrics and opto-electronic materials have been discussed in detail.

This book has been the outcome of endless efforts put in by authors and researchers on various issues and topics within the field. The book is a comprehensive collection of significant researches that are addressed in a variety of chapters. It will surely enhance the knowledge of the field among readers across the globe.

It is indeed an immense pleasure to thank our researchers and authors for their efforts to submit their piece of writing before the deadlines. Finally in the end, I would like to thank my family and colleagues who have been a great source of inspiration and support.

Editor

Part 1

Conventional Sintering

Two-Step Sintering Applied to Ceramics

Gislâine Bezerra Pinto Ferreira, José Ferreira da Silva Jr,
Rubens Maribondo do Nascimento, Uílame Umbelino Gomes
and Antonio Eduardo Martinelli
Federal University of Rio Grande do Norte
Brazil

1. Introduction

During the process of sintering of ceramics, it is necessary to apply high temperature owing the high melting point of the raw materials. In general, a ceramist, wishing to produce a material with particular properties, must identify the required microstructure and then design processing conditions that will produce this required microstructure (Lutgard et al., 2003). One of the options to adapt the microstructure is a technique called two step sintering (TSS), this technique has been applied to the sintering of ceramic oxides to achieve full density without grain growth in final stage of sintering without loss densification (Chen & Wand, 2000). The two-step sintering process consists in to heat a ceramic body to a peak temperature (T1) to achieve an intermediate density and then the temperature is reduced to a dwell temperature (T2), which is held till full density is achieved. To succeed in two-step sintering, a sufficiently high relative density (70% or greater) needs to be achieved at T1 (Chen & Wang, 2000 & Chen, 2000). Once this critical density is reached, a lower temperature, T2, used for the isothermal hold will be sufficient to achieve full density. Difference between kinetics of grain boundary diffusion and grain boundary migration is used to obtain almost full dense, nanostructured ceramics. During the last stage of the sintering occur the grain growth in materials, this implicate in final properties, like mechanical resistance, density, ionic and electrical conductivity and others (Robert et al., 2003). The two-step sintering has been applied in many mateirals with the main goal of avoiding the grain growth in final stage of sintering, the results show the TSS is a technique efficient for it. Some application for two-step sintering are materials which need high density and small grain size, for example electrolytes of solid oxide fuel cell, as ceramics based in Y_2O_3 and CeO_2, both with and without doppant (Wang et al., 2006; Wright, 2008 & Lapa, 2009). Others examples in which TSS are used also as nanostructural fosterite (Fathi, et al., 2009), alumina-zirconia ceramics (Wang et al., 2008), TiBaO3 and Ni-Cu-Zn Ferrite (Wang et al., 2006), ZnO (Shahraki et al., 2010). In this cases, the researchs are getting the relative density higher than 97% and the size grains in level sub-micrometer.

2. Mechanisms of two-step sintering

During the process of the TSS the first step needs high temperature enough to achieve the critical diameter spherical (d_c) of the core to become the crystallization, in this step the relative density (ρ) need be the same or higher than 75% of the teorical density to obtain unstable pores and the sintering the material be kept. In the second step is necessary the temperature to keep the densification until the end of the sintering, but avoiding the grain growth. An important reason to have an understanding of two-step sintering is the possibility to increase the heat rate of sintering, to avoid the grain growth and to obtain a material with improved mechanical, thermal, electrical and optical properties in the materials (Chen & Wang, 2000). The absence of grain growth in second-step sintering has important implications for kinetics. Grain coarsening creates a powerful dynamic that constantly refreshes the microstructure. Statistically, only one-eighth of all grains survive every time the size of the grains doubles. This evolution can be a source of enhanced kinetics. Even without grain growth, enhanced kinetics has also been suspected in cases when microstructure evolution is otherwise robust; for example, in fine-grain superplasticity (McFadden et al., 1999; Wakai et al., 1990 & Chen & Xue, 1990). Because second-step sintering proceeds in a 'frozen' microstructure, it should have slower kinetics. Yet the slower kinetics is sufficient for reaching full density, while providing the benefit of suppressing grain growth. The diffusion kinetics is quantified in the frozen microstructure by measuring the densification rates in the second step, and comparing them with the prediction based on Herring's dimensional argument (Herring, 1950) for normalized densification rate (dr/rdt, where r is relative density and t is time):

$$\frac{d\rho}{\rho dt} = F(\rho)\left(\frac{\gamma\Omega}{GkT}\right)\left(\frac{\delta D}{G^3}\right) \tag{1}$$

Here γ is surface energy, Ω is atomic volume, G is grain size, d is grain-boundary thickness, and D is grain-boundary diffusivity. In the above, $\gamma\Omega/GkT$ may be viewed as the normalized driving force, and $\delta D/G^3$ is the standard kinetic factor that enters the strain-rate equation for grain-boundary processes such as sintering, diffusional creep and superplasticity. The remaining dimensionless prefactor on the right-hand side, F, is independent of the grain size as such, but could depend on other aspects of the microstructure such as density and pore distribution (Wei & Wang, 2000).

Grain boundaries in ceramics have been extensively investigated in recent years with the intent to understand their structures and mechanical/electrical properties. Grain boundaries are also important for kinetic phenomena, such as sintering, grain growth, diffusional creep and superplasticity. Their importance increases as the grain size decreases, since the ratio of grain boundary to the grain interior is inversely proportional to the grain size. In addition, this ratio also dictates that there is a large capillary pressure (and its variation) in fine grain materials. For a typical grain boundary energy (and surface energy) in ceramics of 1 J/m², the capillary pressure is of the order of 2000 MPa at a grain size of 1 nm, 200 MPa at 10 nm, and 20MPa at 100 nm. These pressures are rather significant and may cause additional kinetic effects at intermediate and high temperatures. In general, the dominant kinetic paths in submicron powders and ceramic bodies are surfaces and grain boundaries, the latter becoming increasingly important as the relative density reaches toward 100%. "Clean" experiments on grain boundary kinetics without the "contamination" of surface effects can be undertaken provided full density is first achieved. Common ceramic firing processes, however, always induce rapid grain growth when the relative density exceeds 85%, because of the breakdown of pore channels at three grain junctions and the resulting reduction of the pore drag on grain boundary migration. Nevertheless, the combination of good powder processing, fine starting powders, and low sintering temperatures may help to achieve submicron grain sizes in fully dense bodies. Such ceramics are suitable for kinetic studies of grain boundaries (Chen, 2000). The feasibility of densification without grain growth relies on the suppression of grain-boundary migration while keeping grain boundary diffusion active. Two-step sintering can be used to achieve a relative density of 98% by exploiting the "kinetic window" that separates grain-boundary diffusion and grain-boundary migration. When conditions for two-step sintering fall below the "kinetic window," a density _ 96% cannot be achieved even if a starting density of 70% is achieved at T1, as grain growth may still be suppressed but densification will be exhausted. Above the "kinetic window," grain growth is likely to occur (Wright, 2008). The suppression of the final-stage grain growth is achieved by exploiting the difference in kinetics between grain boundary diffusion and grain-boundary migration. Such a process should facilitate the cost-effective preparation of other nanocrystalline materials for practical applications. To succeed in two-step sintering, a suficiently high starting density should be obtained during the first step. When the density is above 70%, porosimetry data have shown that all pores in Y_2O_3 become subcritical and unstable against shrinkage (which occurs by capillary action). These pores can be called as long as grain-boundary diffusion allows it, even if the particle network is frozen as it clearly is in the second step (Chen & Wang, 2000). From the thermodynamics aspect, at a temperature range where grain boundary diffusion is active, but grain boundary migration is sufficiently sluggish, densification would continue without any significant grain growth. On basis of this idea was developed to suppress the accelerated grain growth at the final stage of sintering by triple junctions. To take the advantage of boundary dragging by triple junctions, a critical density at first should be achieved where sufficient triple junctions exist throughout the body as pins. Then with decreasing the sintering temperature to a critical degree, the grain growth would be stopped by triple junctions while densification may not be impaired. In doing so, samples have to be exposed to prolonged isothermal heating at the second (low temperature) step. As in a TSS regime, the triple junctions are going to prohibit grain growth, while unstable pores can shrink with low temperature annealing, seemingly the source of different densities lies in the pore size and distribution which needs to be further investigated. Certainly, formation of inhomogeneous porosity due to the increased tendency of nanopowder to form agglomerates complicates the situation. To solve this problem, one can use larger particles with lower agglomeration degree and shape green bodies with advanced methods to obtain a more homogenous structure. Under this condition, one can expect successful TSS at lower temperatures (Hesabi et al., 2008).

Sophisticated firing profiles are an alternative to compositional effects, to obtain dense ceramics with proper microstructure. Two-step sintering profiles, including optimized combinations of peak and dwell sintering temperatures, produced nanostructured materials with high densification at reasonably low temperatures, due to different grain growth and densification kinetics (Chen & Wang, 2000). Was observed that samples processed by two-step sintering show best results the density, around 94% of theoretical density and sub-micrometer grain size (Lapa, 2009).

3. Grain boundary kinetics during intermediate and final stage sintering

Sintering data are often used to infer the rate-controlling mechanism following the scaling analysis of Herring (Herring, 1950 & Herring, 1951).

$$\frac{d\rho}{dt} = \left(\frac{f(\rho)}{kTR^m}\right) D_0 \, exp(-Q/kT) \tag{2}$$

In the above, ρ is the relative density, m is either 3 for lattice diffusion or 4 for grain boundary diffusion, D_0 and Q refer to the pre-factor and activation energy of either lattice diffusion or grain boundary diffusion, and f is a constant that is dependent on the pore/grain geometry of the sintering body. Over a range of relative density, from 60% to 90%, some model calculations suggest that f is relatively constant (Coble, 1965 & Swinkels and Ashby, 1981). Thus, Eq. (2) can be used to deduce the diffusivity and the rate controlling mechanism if the densification rate and the grain size are known. In practice, plotting Log $(Td\rho=dt)R^m$ against $1/T$ usually yields a straight line regardless whether m is chosen as 3 or 4. This is due to the relatively poor resolution of such plotting method, the unavoidable scatter of the data, and the uncertainty of the value of $f(\rho)$. Therefore, it is usually not possible to definitively state that the sintering mechanism is via grain boundary diffusion or lattice diffusion based on the scaling analysis alone. On the other hand, the inferred values of diffusivities, and especially those of the activation energy, are usually quite different depending on whether m is chosen as 3 or 4. Thus, when independent diffusivity data are available, e.g., from grain boundary mobility measurements, a self-consistency check may be applied to infer whether the lattice or grain boundary mechanism applies. Using this method, is possible to conclude that later stage sintering of submicron CeO_2 and Y_2O_3 powders is controlled by grain boundary diffusion (Chen & Chen, 1997). As cited above, to achieve densification without grain growth, is necessary to first fire the ceramic at a higher temperature ($T1$) to a relative density of 75% or more, then sinter it at a lower temperature ($T2$) for an extended time to reach full density. This schedule is different from the conventional practice for sintering ceramics, in which the temperature always increases or, at least is held constant at the highest temperature, until densification is complete. The temperature $T2$ required for the second step decreased with the increasing grain size. However, if $T2$ is too low, then sintering proceeds for a while and then becomes exhausted. On the other hand, if $T2$ is too high, grain growth still occurs in the second step. Chen 2000, observed that the same shape in different Y_2O_3 but the medium temperature is shifted depending on the solute added. The results imply that grain boundaries previously stabilized at a higher temperature are difficult to migrate at a lower temperature even though they may still provide fast diffusion paths. The presence of impurity or solute segregation is not essential for this observation, since the same observation was found in pure Y_2O_3 as well as Y_2O_3 doped with both diffusion enhancing and diffusion-suppressing solutes. Therefore, the suppression of migration is not due to solute pinning. This suggests, for the first time that the mechanism for grain boundary migration is not grain boundary diffusion even in a pure substance. If the activation energy of the additional step is higher than that of grain boundary diffusion, it could explain why grain boundary migration is inhibited at low temperatures but not at high temperature. The most likely candidates for such step are movement of nodal points or nodal lines on the grain boundary, such as four-grain junctions, pore-grain boundary junctions, or three-grain junctions. The structures of these nodal points and lines may be special and they could become stabilized by prior high temperature treatment, rendering them difficult to alter to accommodate the subsequent movement of migrating grain boundary at low temperatures. Empirically, this may be modeled by assigning mobility to the nodal point (line), whose ratio to grain boundary mobility decreases with increasing temperature.

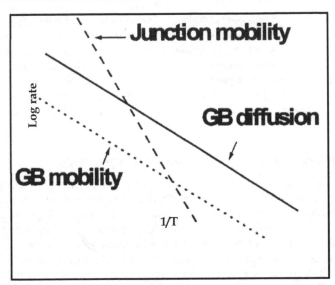

Fig. 1. Schematic Arrhenius plot for grain boundary diffusion, mobility of pore/grain-boundary junction or four-grain junction, and intrinsic mobility of grain boundary (without extrinsic drag due to nodal points/lines) (Source: Gary J. Wright, 2008).

Simple arguments then show that below a certain temperature, equilibrium grain boundary migration does not obtain since the boundaries are effectively pinned by the nodal points (lines). As mentioned above, enhanced grain boundary migration is often observed in superplastic deformation of fine grain oxides. The grain growth in this case is found to be controlled by the plastic strain. Indeed the ratio of grain size is essentially of the same order as the ratio of specimen dimensions before and after deformation (Chen & Xue, 1990). This may be regarded as opposite to the suppression of grain boundary migration described above. It is likely that in both cases, the dynamics of the nodal line/point are important. In superplasticity, the dynamics are enhanced to facilitate grain boundary migration. In low temperature sintering, the dynamics are inhibited to suppress grain boundary migration. A better knowledge of the structures of the grain boundary nodal points and lines, in both equilibrium configurations and in dynamic configurations would be required for a full understanding of the grain boundary kinetics. One interesting observation though is that a parallel effect of solute is seen in all three cases: normal grain growth, dynamic grain growth, and sintering without grain growth. For example, solutes that enhance normal grain growth also cause faster dynamic grain growth, and solutes that suppress normal grain growth likewise show a higher temperature $T2$ in the kinetic window for sintering without grain growth. Thus, while the kinetics of the nodal point/line may be distinct from that of grain boundary diffusion, they may not be entirely independent of each other. Recent studies of high-purity zinc have shown that grain-boundary migration can be severely hampered by the slow mobility of grain junctions at lower temperatures, the latter having a higher activation energy (Czubayko, et al., 1998). It is possible that a similar process, in which grain junctions as well as grain boundary/pore junctions impede grain-boundary migration, may here explain the apparent suppression of grain growth at lower temperatures. Interface kinetics in very fine grain polycrystals is sometimes limited due to difficulties in maintaining sources and sinks to accommodate point defects 21±23. This leads to a threshold energy or stress, of the order of 2g/G. For a grain size of 100 nm, this amount to 20MPa, which is rather substantial compared to capillary pressure and could be the cause for the suppression. This effect should diminish at larger grain sizes, allowing the kinetic window to extend to lower temperatures. Therefore, in exploiting the difference in the kinetics of grain-boundary diffusion and grain-boundary migration to achieve densification without growth at lower temperatures, it is still advisable to utilize dopants to 'tune' the overall kinetics.

4. Influence of triple junctions on grain boundary motion

According Czubayco et al., during the formation of granular structure of a polycrystalline material, both grain boundaries and triple junctions influence in characteristics. In the past, just the of grain boundaries motion were studied, while the influence of triple junctions on grain boundary has not attention necessary. In the last years, the velocity of the junction motion, the shape of the intersecting grain boundaries must be measured. Moreover, the steady-state motion of a grain boundary system with a triple junction is only possible in a very narrow range of geometrical boundary configurations. It is usually assumed that triple junctions do not influence the motion of the adjoining grain boundaries and that their role is reduced to control the thermodynamic equilibrium angles at the junction during the boundary motion. A specific mobility of triple junctions was first introduced by Shvindlerman and co-workers (Galina et al., 1987), who considered the steady-state motion of a grain boundary system with a triple junction. The geometry of the used boundary system is shown in Fig. 2. The boundaries of this system are perpendicular to the plane of the diagram, and far from the triple junction they run parallel to one another and to the x-axis.

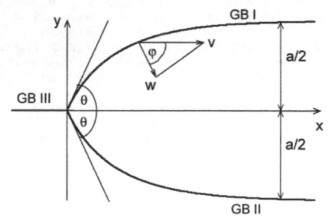

Fig. 2. Geometry of the grain boundary system with triple junction in the course of steady state motion (From: Czubayko et al., 1998).

The three boundaries of the system are considered identical, in particular their surface tension σ and their mobility m^{GB}. Furthermore, it is assumed that m^{GB} and σ are independent of the inclination of the grain boundaries. These assumptions define the problem to be symmetric with regard to the x-axis. With these simplifications some very important features of the motion of this system can be established. (a) A steady-state motion of the whole system is possible indeed. (b) The dimensionless criterion Λ:

$$\Lambda = \frac{m^{TJ}a}{m^{GB}} = \frac{2\theta}{2\cos\theta - 1} \tag{3}$$

describes the drag influence of the triple junction on the motion of the entire boundary system. For $\Lambda \gg 1$ the junction does not drag the motion of the boundary system, and the angle θ tends to the equilibrium value $\pi/3$. In such a case the velocity v of the motion of the entire boundary system is independent of the mobility of the triple junction and is determined by the grain boundary mobility and the acting driving force:

$$v = \frac{2\pi m^{GB}\sigma}{3a} \tag{4}$$

In contrast, for $\Lambda \ll 1$ the motion of the system is controlled by the motion of the triple junction and the angle θ tends to zero. The velocity depends only on the triple junction mobility and the grain boundary surface tension σ:

$$v = \sigma m^{TJ} \tag{5}$$

Owing to the fact that there are no measurements and no data of triple junction mobility, we cannot even estimate whether the ratio m^{TJ}/m^{GB} is finite. On the other hand, in the course of triple junction motion the straight grain boundary (Fig. 2, GB III) has to be extended. The velocity of its formation is unknown, but the kinetics of it should depend on the structure and properties of the generated grain boundary. Insofar as the rate of formation of this boundary can be interpreted as the velocity of the triple junction, which is proportional to its mobility according to equation (5). In the following a boundary system as shown in Fig. 2 with two identical curved boundaries (GB I and II) and a di€erent straight boundary (GB III) will be considered. The respective surface tensions and the mobilities of the boundaries are:

$$\sigma 1 = \sigma 2 \equiv \sigma \neq \sigma 3, m_1^{GB} = m_2^{GB} \equiv m^{GB} \neq m_3^{GB} \tag{6}$$

In this case the shape of a steadily moving boundary system can be expressed by the equation

$$\frac{d^2y}{dx^2} = -\frac{v}{m^{GB}\sigma}\frac{dy}{dx}\left[1 + \left(\frac{dy}{dx}\right)^2\right] \tag{7}$$

with the boundary conditions

$$y(0) = 0, y(\infty) = \frac{a}{2}, y'(0) = \tan\theta \tag{8}$$

as obvious from Fig. 2. Equations (6)-(8) completely define the problem. The shape of the stationary moving grain boundaries GB I and II (Fig. 1) is given by:

$$y(x) = \xi arccos\left(e^{-\frac{x}{\xi}+C_1}\right) + C_2, \ \xi = \frac{a}{2\theta} \tag{9}$$

$$C_1 = \frac{1}{2}ln(\sin\theta)^2, C_2 = \xi\left(\frac{\pi}{2} - \theta\right) \tag{10}$$

The steady-state velocity of GB I and II is

$$v^{GB} = \frac{2\theta m^{GB}\sigma}{a} \tag{11}$$

The velocity of the triple junction v^{TJ} can be expressed as (Soraes, et al., 1941), (Galina, et al., 1987), (Fradkov, et al., 1988):

$$v^{TJ} = m^{TJ} \sum \sigma_i \vec{\tau_i} \tag{12}$$

where every $\vec{\tau_i}$ is the unit vector normal to the triple line and aligned with the plane of the adjacent boundary. If the angles at the triple junction are in equilibrium, the driving force is equal to zero and for a finite triple junction mobility the velocity v^{TJ} should vanish as well. Consequently, for a finite m^{TJ}, the motion of the triple junction disturbs the equilibrium of the angles and, as a result, drags the motion of the boundaries. For the situation given in Fig. 2.

$$v^{TJ} = m^{TJ}(2\sigma \cos\theta - \sigma_3) \tag{13}$$

In the case of steady-state motion of the entire boundary system the velocity of the triple junction equals the velocity of the grain boundaries. Therefore, the steady-state value of the angle θ is determined by equations (11) and (12):

$$\frac{2\theta}{2\cos\theta - \frac{\sigma_3}{\sigma}} = \frac{m^{TJ}a}{m^{GB}} = \Lambda \tag{14}$$

The dimensionless criterion Λ reflects the drag influence of the triple junction on the migration of the system. One can distinguish two limiting cases:

$\Lambda \to 0$: In this case the angle θ tends to zero, i.e. the motion of the entire boundary system is governed by the mobility of the triple junction and the corresponding driving force. For the limit $\theta = 0°$ the velocity of the system is given with equation (13) by

$$v = m^{TJ}(2\sigma - \sigma_3) \tag{15}$$

$\Lambda \to \infty$: In this case the angle θ tends to the value of thermodynamic equilibrium:

$$\theta = arccos\left(\frac{\sigma_3}{2\sigma}\right) = \theta_{eq.} \tag{16}$$

The motion of the system is independent of the triple junction mobility and is governed only by the grain boundary mobility and the corresponding driving force. The velocity of the boundary system in this case with equations (11) and (16) is given by:

$$v = \frac{2\theta_{eq} m_{GB}\sigma}{a} \tag{17}$$

The two states of motion of the entire grain boundary system can be distinguished experimentally for a known ratio $\sigma3/\sigma$ by measuring the contact angle θ.

5. Influence of external shear stresses on grain boundary migration

A method to activate and investigate the migration of planar, symmetrical tilt boundaries is influenced by external shear stress. It is shown that low- as well as high-angle boundaries could be moved by this shear stress. From the activation parameters for grain boundary migration, the transition from low- to high-angle boundaries can be determined. The migration kinetics were compared with results on curved boundaries, and it was shown that the kinetics of stress induced motion were different from the migration kinetics of curvature driven boundaries. Washburn, et al. 1952 and Li, et al., 1953 investigated planar low-angle boundaries in Zn under the influence of an external shear stress and observed the motion with polarized light in an optical microscope. Symmetrical low angle tilt boundaries consist of periodic arrangements of a single sets of edge dislocations. An external shear stress perpendicular to the boundary plane will cause a force on each dislocation and in summary a driving force on the boundary. The samples were exposed to a shear stress ranging from 10^{-1} to 10^{-3}MPa. In aluminum (purity 99.999%) the yield stress is 15–20MPa, hence the applied shear stress is definitely in the elastic range. High angle symmetrical tilt boundaries also can be formally described as an arrangement of a single set of edge dislocations except that the dislocation cores overlap and the identity of the dislocations gets lost in the relaxed boundary structure. I showed that irrespective of the magnitude of the angle of rotation, grain boundaries can be moved under the action of the applied shear stress. The transition from low- to high-angle grain boundaries is revealed by a conspicuous step of the activation enthalpy at a misorientation angle of 13.6°. This holds for low angle as well as for high angle symmetrical tilt boundaries. For the curvature driven grain boundaries our results are in good agreement with previous experimental data [14] and one can see a strong dependency of the activation enthalpy on the misorientation angle, i.e. on the grain boundary structure. There is also a clear difference between the activation enthalpies for the stress induced motion of the planar high angle grain boundaries and the curvature driven migration of the curved high angle grain boundaries. Obviously, a dislocation in a high angle grain boundary does not relax completely its strain field and correspondingly, a biased elastic energy density induced by an applied shear stress will induce a force on all dislocations that comprise the grain boundary. The results prove that grain boundaries can be driven by an applied shear stress irrespective whether low- or high-angle boundaries. Obviously, the motion of the grain boundary is caused by the movement of the dislocations, which compose the grain boundary. The motion of an edge dislocation in a FCC crystal in reaction to an applied shear stress ought to be purely mechanical and not thermally activated. Obviously, the observed grain boundary motion is a thermally activated process controlled by diffusion. To understand this, one has to recognize first that grain boundary motion is a drift motion since it experiences a driving force that is smaller compared with thermal energy. Moreover, real boundaries are never perfect symmetrical tilt boundaries but always contain structural dislocations of other Burgers vectors. These dislocations have to be displaced by nonconservative motion to make the entire boundary migrate. The climb process requires diffusion, which can only be volume diffusion for low angle grain boundaries but grain boundary diffusion for high angle grain boundaries according to the observed activation enthalpies. The different behavior of curvature driven grain boundaries is not due to the curvature of the boundaries rather than due to a different effect of the respective driving force. While an applied shear stress couples with the dislocation content of the boundary in a curved grain boundary each individual atom experiences a drift pressure to move in order to reduce curvature.

6. Shape of the moving grain boundaries

The principal parameter which controls the motion of a grain boundary is the grain boundary mobility. In practically all relevant cases the motion of a straight grain boundary is the exception rather than the rule. That is why the shape of a moving grain boundary is of interest and it will be shown that the grain boundary shape is a source of new interesting and useful findings of grain boundary motion, for the interaction of a moving grain boundary with mobile particles, in particular. The experimentally derived shape of grain boundary 'quarter-loop' in Al-bicrystals of different purity was compared with theoretical calculations in the Lücke–Detert approximation. The shape of a moving GB quarter-loop was determined analytically under the assumption of uniform GB properties and quasi-two-dimensionality (Verhasselt, et al., 1999):

$$y(x) = \{-(b_F - b_L)arccos\left(\frac{\sin\Theta}{e^{x^*/b_F}}\right) + \frac{a}{2} - b_L\frac{\pi}{2} + b_F arccos\left(e^{b_F\ln(\sin\Theta)-x/b_F}\right)$$

$$0 \leq x \leq x^*\frac{a}{2} - b_L\frac{\pi}{2} + b_L arccos\left(e^{b_L\ln(\sin\Theta)-x^*((b_L/b_F)-1)-x/b_L}\right) \quad (18)$$

$$x \geq x^*$$

where b_L is:

$$b_L = \frac{b_F\left(arc\cos(\sin\Theta/e^{x^*/b_F}+\Theta-(\pi/2)-a/2)\right)}{arc\cos(\sin\Theta/e^{x^*/b_F})-(\pi/2)}b \quad (19)$$

The parameters in Eq. (18) are the width of the shrinking grain $a/2$, the angle Θ of the grain boundary with the free surfaces in the triple junction, the critical point x^*, and b_F and b_L: $b_L=m_L\sigma/V$; $b_F=m_F\sigma/V$, where m_L and m_F are the GB mobilities for 'loaded' and 'free' GB, respectively, σ is GB surface tension, V is a velocity of a quarter-loop. The first two parameters can be measured directly in the experiment. The latter two have to be chosen in an approximate way to fit the experimentally derived grain boundary shape. The point x^* is the point of intersection 'free' and 'loaded' segments of the GB. The value m_L/m_F is a measure for how drastic the change between the 'free' and the 'loaded' part in the point of intersection will be. The investigation proves that the influence of the impurity atoms on grain boundary properties and behavior is rather strong even in very pure materials. As mentioned above, the shape of a moving grain boundary is a new source of information on grain boundary migration. One example is given in Fig. 4, where the value of the critical distance x^*, normalized by the driving force (in terms of the quarter-loop width a) is plotted versus the impurity content. In accordance with the Lücke–Detert theory the critical velocity v^* (and rigidly bound to it the position of the critical point x^* on the quarter-loop) is determined by the balance between the maximum force of interaction of the impurity atoms with the boundary and the force, which is imposed by the energy dissipation caused by boundary motion across the matrix. The difference of the impurity drag for grain boundaries in samples with different amount of impurities is caused by the adsorption of impurities at the grain boundary. According to theory, the velocity should decrease proportionally to the inverse of the concentration of adsorbed atoms. Therefore x^* should increase with decreasing impurity content, as observed qualitatively (Fig. 4 and Fig. 5). However, a linear relation between the inverse of the impurity concentration and v^*, i.e. x^*, is not observed over the whole concentration range, which indicates a more complicated interaction of adsorbed atoms with the grain boundary. In such a case, x^*/a should increase more strongly with decreasing impurity content than it does linearly. This tendency is indeed observed (Fig. 4 and Fig. 5).

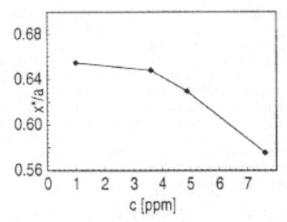

Fig. 4. Dependence of critical point x^*/a on impurity content; (b) reciprocal impurity content (From: Shvindlerman, Gottstein, 2001).

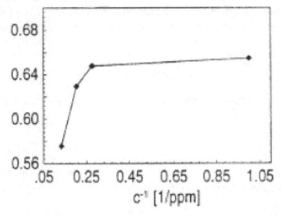

Fig. 5. Dependence of critical point x^*/a on reciprocal impurity content (From: Shvindlerman, Gottstein, 2001).

7. Dragging effect of tripe junction on grain boundary motion

In spite of the fact that a line (or column) of intersection of three boundaries constitutes a system with specific thermodynamic properties was realized more than 100 years ago (by Gibbs), the kinetic properties of this subject, in particular the mobility of triple junctions, and their influence on grain growth and relevant processes were ignored up to now. Although the number of triple junctions in polycrystals is comparable in magnitude with the number of boundaries, all peculiarities in the behavior of polycrystals during grain growth were solely attributed to the motion of grain boundaries so far. It was tacitly assumed in theoretical approaches, computer simulations and interpretation of experiment results that triple junctions do not disturb grain boundary motion and that their role in grain growth is reduced to preserve the thermodynamically prescribed equilibrium angles at the lines (or the points for 2-D systems) where boundaries meet. The most prominent example of how this assumption determines the fundamental concepts of grain structure evolution gives the Von Neumann–Mullins relation (Neumann, 1952 and Mullins, 1956). No doubt this relation forms the basis for practically all theoretical and experimental investigations as well as computer simulations of microstructure evolution in 2-D polycrystals in the course of grain growth. This relation is based on three essential assumptions, namely, (i) all grain boundaries possess equal mobilities and surface tensions, irrespective of their misorientation and crystallographic orientation of the boundaries; (ii) the mobility of a grain boundary is independent of its velocity; (iii) the third assumption relates directly to the triple junctions, namely, they do not affect grain boundary motion; therefore, the contact angles at triple junctions are in equilibrium and, due to the first assumption, are equal to 120°. As it was shown in (Neumann, 1952 and Mullins, 1956), for 2-D grain, the rate of change of the grain area S can be expressed by:

$$\frac{dS}{dt} = -A_b \oint d\varphi \tag{20}$$

where $A_b = m_b \sigma$; m_b being the grain boundary mobility, σ is the grain boundary surface tension. If the grain were bordered by a smooth line, the integral in Eq. (20) would equal 2π. However, owing to the discontinuous angular change at every triple junction, the angular interval $\Delta\varphi = \pi/3$ is subtracted from the total value of 2π for each triple junction. Consequently:

$$\frac{dS}{dt} = -A_b \left(2\pi - \frac{n\pi}{3}\right) = \frac{A_b \pi}{3}(n-6) \tag{21}$$

where n is the number of triple junctions for each respective grain, i.e. the topological class of the grain.

8. Conclusions

Many researchers have used the two-step sintering as a design process to obtain samples with a microstcture without grain growth in final stage of sintering. Same exemples that we can cite are:

- Chen, I.W. & Wang, X.H. (2000) obtained samples of the Y_2O_3 with a grain size of 60nm can be prepared by a simple two-step sintering method, at temperatures of about 1,000°C without applied pressure. The suppression of the final-stage grain growth is achieved by exploiting the difference in kinetics between grain boundary diffusion and grain-boundary migration. Such a process should facilitate the cost-effective preparation of other nanocrystalline materials for practical applications.
- 2: Lapa, et al., (2009) prepared samples of the yttrium and gadolinium-doped ceria-based electrolytes (20 at% dopant cation) with and without small Ga_2O_3-additions (0.5 mol%). The average grain sizes in the range 150–250 nm and densifications up to about 94% were found dependent on the sintering profile and presence of Ga. The grain boundary arcs in the impedance spectra increased significantly with Ga-doping, cancelling the apparently positive role of Ga on bulk transport, evidenced mostly in the case of yttrium-doped materials.
- 3: Wang, et al. (2006) used two-step sintering to sinter $BaTiO_3$ and Ni–Cu–Zn ferrite ceramics to high density with unprecedentedly fine grain size, by suppressing grain

growth in the final stage of densification. Dense $BaTiO_3$ ceramics with a grain size of 35 nm undergo distortions from cubic to various low-temperature ferroelectric structures. Dense fine grain Ni–Cu–Zn ferrite ceramics have the same saturation magnetization as their coarse grain counterparts.

9. Acknowledgment

The authors wish to thank PRH-ANP, CAPES, LMCME-UFRN, Materials Laboratory-UFRN and NEPGN-UFRN.

10. References

Arzt, E., Ashby, M. F. & Verrall, R. A. (1993) Interface-controlled diffusional creep. Acta Metall. 31, 1977± 1989.

Cannon, R. M., Rhodes, W. H. & Heuer, A. H. (1980) Plastic deformation of fine-grained alumina: I. interface-controlled diffusional creep. J. Am. Ceram. Soc. 63, 48-53

Chen, I.W. (1993). Mobility control of ceramic grain boundaries and interfaces, Materials Science and Engineering. 166. 51-58.

Chen, I.W. (2000). Grain boundary kinectics in oxide ceramics with the cubic fluorite crystal structure and its derivates. Interface Science. Vol. 8. Pp. 147-156.

Chen, I.W. & Wang, X.H. (2000). Sintering dense nanocrystaline ceramics without final-stage grain growth. Nature. Vol. 404.

Chen, I.W. & Xue, L.A. (1990). Development of superplastic structure ceramics. Journal of American Ceramics Society. Vol. 73.

Chen, P.L. & Chen, I.W. (1997). Journal of American Ceramics Society.

Coble, R.L. (1965). Intermediate-stage sintering : Modification and correction of a lattice-diffusion model. Vol. 36.

Czubayko, L., Sursaeva, V. G., Gottstein, G. & Shvindlerman, L. S., (1998) Influence of triple junctions on grain boundary motion. Acta Mater. 46, 5863±5871.

Galina, A. V., Fradkov, V. E. and Shvindlerman, L. S., (1987) Physics Metals Metallogr., 63, 165.

Herring, C. (1950). Effect of change of scale on sintering phenomena. Journal Applied to Physics. Vol. 21.

Herring, C. (1951). The physics of powder metallurgy. McGraw-Hill. New York. Pp. 143.

Hesabi, Z.R. ; Haghighatzadeh, M. ; Mazaheri, M. ; Galusek, D.S.K. & Sadrnezhaad. (2008) Suppression of grain growth in sub-micrometer alumina via two-step sintering method. Journal of the European Ceramic Society.

Jonghe, L.C. ; Rahaman, M.N. (2003). Sinterig ceramics. Handbook of Advanced Ceramics.

Land, T. A., Martin, T. L., Potapenko, S., Palmore, G. T. & De Yoreo, J. J. (1999) Recovery of surfaces from impurity poisoning during crystal growth. Nature 399, 442±445.

Lapa, C.M. ; Souza, D.P. ; Figueiredo, F.M.L. & Marques, F.M.B. (2009). Electrical and microstructural characterization of two-step sintered ceria based electrolytes. Journal of Power Sources. Vol. 187. Pp. 204-208.

Lapa, C.M. ; Souza, D.P. ; Figueiredo, F.M.L. & Marques, F.M.B. (2009). Two-step sintering ceria-based electrolytes. International Journal of Hydrogen Energy. Pp. 1-5.

Li, C.H., Edwards, E.H., Washburn, J., Parker, E.R., (1954) Recent observations on the motion of small angle dislocation boundaries   Acta Met. 2, 322–333.

Mullins, W.W.,(1956), Two-dimensional motion of idealized grain boundaries. J. Appl. Phys. 27, 900–904.

Fathi, M.H.; Kharaziha, M. (2009). Two-step sintering of dense, nanostructural forsterite. Materials Letters - 63, Pp. 1455–1458.

Fradkov, V. E. and Shvindlerman, L. S., (1988) Structure and Properties of Interfaces in Metals. Nauka, Moscow, p. 213.

McFadden, S.X. ; Mishra, R.S. ; Valiev, R.Z. ; Zhilyaev, A.P. & Mukherjee, A.K. (1999). Low-temperature superplasticity in nanostructured nickel and metal alloys. Nature. Vol. 396. Pp. 684-686.

Robert, C.L. ; Ansart, F. ; Degolet, C.L. ; Gaudon, M. & Rousset, A. (2003). Dense yttria stabilized zirconia: sintering and microstructure Ceramics International. Vol. 29. Pp. 151-158.

Shahraki, M.M.; Shojaee, S.A.; Sani, M.A.D.; Nemati, A. & Safaee, I. (2010). Two-step sintering of ZnO varistors. Solid State Ionics.

Shvindlerman, L.S., Gottstein, G., (2001), Grain boundary and triple junction migration Materials Science and Engineering, A302, 141–150.

Soraes, A., Ferro, A. C. and Fortes, M. A., (1941) Scripta metallurgycal., 1985, 19.

Swinkels, F.B. & Ashby M.F. (1981). Acta Metallurgycal. Vol. 29. Pp. 259.

Verhasselt, J.C., Gottstein, G., Molodov, D.A., Shvindlerman, L.S., (1999) Acta Mater. 47, 887–892.

Von Neumann, J., (1952) in: Metal Interfaces, American Society for Testing Materials, Cleveland, OH, P. 108.

Wakai, F. et al. (1990) A superplastic covalent crystal composite. Nature. Vol. 344.

Wang, J.C.; Huang, C.Y.; Wu, Y.C. (2008) Two-step sintering of fine alumina-zirconia ceramics. Ceramics international.

Wang, X.H. ; Chen, P.L. & Chen, I.W. (2006). Two-step sintering of ceramics with constant grain-size, I Y_2O_3. Journal American Ceramics Society. Vol. 89. Pp. 431-437.

Wang, X.H.; Deng, X.Y.; Bai, H.L.; Zhou H.; Qu, W.G. & Li, L.T. (2006). Two-step sintering of ceramics with constant grain-size, II $BaTiO3$ and Ni-Cu-Zn ferrite. Journal American Ceramics Society.Vol. 89. 438-443.

Washburn, J., Parker,E.R., J. (1952), Journal of Metals. 4, 1076–1078.

Winning, M., Gottstein,G., Shvindlerman, L.S., in: T. Sakai, G. Suzuki (Eds.), (1999), Recrystallization and Related Phenomena, The Japan Institute of Metals, pp. 451–456.

Wright, G.J. (2008). Constrained sintering of yttria-stabilized zirconia electrolytes: The influence of two-step sintering profiles on microstructure and gas permeance. International Journal Applied to Ceramic Technologies. Vol. 5. Pp. 589-596.

Mechanisms of Microstructure Control in Conventional Sintering

Adriana Scoton Antonio Chinelatto[1],
Elíria Maria de Jesus Agnolon Pallone[2],
Ana Maria de Souza[1], Milena Kowalczuk Manosso[1],
Adilson Luiz Chinelatto[1] and Roberto Tomasi[3]

[1]*Department of Materials Engineering - State University of Ponta Grossa*
[2]*Department of Basic Sciences – FZEA - São Paulo University*
[3]*Department of Materials Engineering - Federal University of São Carlos*
Brazil

1. Introduction

The manner and mechanisms involved on the sintering process are essential investigation to achieve the required microstructure and final properties in solids. During the conventional sintering of a compacted powder, densification and grain growth occur simultaneously through atomic diffusion mechanisms. Many researchers have been working on reducing the grain size below 1 μm aiming to improve some properties, such as strength, toughness and wear resistance in ceramics (Greer, 1998; Inoue & Masumoto, 1993; Morris, 1998). In order to obtain ultra-fine ceramic microstructures, nanocrystalline powders can be used. Although the sinterability of nanoparticles is superior to that of fine particles due to the higher sintering stress, densification of these powders is often accompanied by grain growth (Suryanarayana, 1995).

Hot pressing sintering (He & Ma, 2000; Porat et al., 1996), spark plasma sintering (Gao et al., 2000; Chakravarty et al., 2008) or pulse electric current sintering (Zhou et al., 2004) are typical techniques employed to produce nanostructured ceramics. However, many of these techniques are not economically viable depending on the use of the final product. Thus, conventional pressureless sintering is still a more attractive sintering method to produce ceramic products, mainly due to its simplicity and cost compared to other methods. In the conventional pressureless sintering, a controlled grain size with high densification could be achieved by adequate control procedures of the heating curve — herein defined as the maximization of the final density with minimum grain growth.

One hypothesis to the heating curve control can be achieved by improving the narrowing of grain size distribution in a pre-densification sintering stage followed by a final densification stage namely at a maximum densification rate temperature (Chu et al., 1991; Lin & DeJonghe, 1997a, 1997b). In a thermodynamics point-of-view, another hypothesis is regarded to control the heating schedule at temperature ranging the active grain boundary diffusion. Note, however, that the grain boundary migration is sufficiently sluggish and the densification could occur without grain growth. The aforementioned hypothesis was proposed by Chen and Wang (Chen & Wang, 2000) and has been successfully applied to different types of materials. A second phase can be added to preserve fine grains. In this case, grain boundary inhibition can be due to the pinning effect, which is associated with particles locations at grain boundaries or triple junctions (Chaim e al. 1998; Trombini et al., 2007). This drag pinning effect associated with heating curve control can be more effective to suppress the grain growth.

2. Nanostructure materials

The size control of microstructural elements has been always considered as one of the most important factors in control of several properties in the development of new materials or design new microstructures. As a historical example, it can be mentioned that the grain refining of metallic materials, which results in increased mechanical strength, tenacity, occurrence of superplastic, etc. New materials with sub-micron grain size have been developed recently as commercial materials and the latest generation of this development is the nanostructured materials (Inoue & Masumoto, 1993).

Nanostructured materials (also called nanocrystalline materials, nanophasics materials or nanometer-sized crystalline solids) are known to have properties or combinations of properties, which may be new or even superior to conventional materials (Greer, 1998). Nanostructured materials can be defined as a system containing at least one microstructural nano characteristic (with sizes ranging up to 100-150nm). Due the extremely small dimensions, a large volume fractions atoms located in grain boundaries (Morris, 1998), which gives them a unique combination of composition and microstructure (Suryanarayana, 1995). Generally, these materials exhibit high strength and hardness, increased diffusivity, improved ductility and toughness, reduced elastic modulus, lower thermal conductivity when compared to larger grain size materials ($\sim 10\mu m$) (Suryanarayana, 1995).

Since nanocrystalline materials contain a large fraction of atoms in grain boundaries, many of these interfaces provide high density of short diffusion paths. Therefore, it is expected that these materials show increased diffusivity compared to polycrystalline materials of the same composition and conventional particle size (of the order of microns) (He & Ma, 2000). The consequence of such increased diffusivity is increased sinterability of nanometric powders, which causes decrease in sintering temperature of these powders when compared to the same material with conventional particle size (Porat et al., 1996).

The driving force for sintering or "sintering stress" of nanocrystalline ceramics with pores in the range of 5nm is about 400MPa (considering γ about $1Jm^{-2}$), while for conventional ceramics with pores around 1 μm it is 2 MPa. Thus, a nanocrystalline ceramic must have a great thermodynamic driving force for retraction, which must densify extremely well even under unfavorable kinetic conditions such as low temperatures (Suryanarayana, 1995).

The interest in this nanostructured materials area has grown due to the availability of nanocrystalline ceramic powders. These nanocrystalline powders can be synthesized using different techniques, but its consolidation into dense ceramics without significant grain growth is still a challenge.

Compaction and sintering of ultra fine and/or nanoscale powder have a positive set of aspects over behavior during processing and final properties of products; however, there are also several processing difficulties. Main positive aspects include: increased reactivity between reagents and solid particles and between particles and the gas phase, which are important processes in synthesis; increased sintering rate and particularly lowering of sintering temperature, which can reduced by half the material's melting point (Hahn, 1993; Mayo, 1996).

On the other hand, also due to the large surface area and large excess of free energy nanometric powder systems, there are many detrimental aspects to the processing and obtaining the refined and homogeneous microstructures. Some of these aspects are: a very strong tendency to agglomeration of primary particles of nanometric powders; difficulties of mixing and homogenization of compression due to the strong attraction between particles; demand for greater sintering atmosphere control, not only due to the higher reactivity, but also the possibility of formation of thermodynamically unstable phases and appearance of strong effect of adsorbed gases on the surface (Allen et al., 1996; Averback et al., 1992).

Many studies (Chen & Chen, 1996, 1997) on nanometric size particles have shown reduction of sintering temperature. Hahn et al. (Hanh, 1990), studying the sintering of nanometric TiO_2 (12nm), Y_2O_3 (4 nm) and ZrO_2 (8nm), found lower sintering temperatures than those conventional. The sintering of TiO_2 occurred at 1000°C while conventional TiO_2 sintering requires temperatures above 1400°C. The same pattern of reduced sintering temperature was observed for Y_2O_3 and ZrO_2. In spite of the proven decrease in sintering temperature of nanometric powders, its densification is often accompanied by a large grain growth, causing lost of their nanocrystalline ceramics characteristics.

2.1 Effect of heating curve in the sintering

Production of polycrystalline ceramics with high density and small grain size have been studied for several processing routes. Among these routes may be cited: colloidal processes of powder with controlled particle size distribution (Sigmund & Bergström, 2000; Lim et al., 1997), sintering under pressure (He & Ma, 2000; Weibel et al.,1997), use of additives incorporated into a second phase or in solid solution (Novkov, 2006; Erkalfa et al., 1996), spark plasma sintering (Gao et al., 2000; Chakravarty et al., 2008, Bernard-Granger & Guizard, 2007), pulse electric current sintering (Zhou et al., 2004), etc. Usually these methods have several limitations on usage, in addition to requiring more complex and expensive equipment. Thus, sintering without pressure is even a more desirable sintering method to produce ceramic products, mainly due to its simplicity and cost when compared to other methods.

In pressureless sintering, beyond the control of powders' characteristics, control of the sintering process has a major effect on final material's density and microstructure. This method is often unable to prepare dense ceramics with ultrafine grain size, once way the final sintering stage, both densification and grain growth occur by the same diffusion mechanisms (Mazahery et al., 2009).

Heating curve control to manipulate the microstructure during sintering is a route that has been studied and offers advantages such as simplicity and economy. The rate-controlled sintering (Brook, 1982, German, 1996) is one of the ways in which the relationship between densification rate and grain growth rate is determined to identify the sintering temperature at which densification rate is maximized (Chu et al., 1991). Ragulya and Skoroklod (Ragulya & Skoroklod, 1995) studied the rate-controlled sintering of ultra fine nickel powders obtaining sintered samples with high densities (~ 99% TD) and grain size smaller than 100nm. Based on their results, they stated that rate-controlled sintering is a possible route for obtaining dense materials with nanocrystalline structure.

A direct consequence of the rate-controlled sintering method is fast firing (Harmer & Brook, 1981), which can produce dense materials with small grain size, minimizing the time of exposure at temperatures where grain growth is fast compared with densification (Chu et al., 1991). This is possible because, generally, coalescence mechanisms (eg, surface diffusion and vapor transport) prevail over densification mechanisms (eg, volumetric diffusion by grain boundary diffusion) at low temperatures. In this case, shorter times at lower temperatures reduce growth, so that the driving force for densification is not decreased significantly (Lin & DeJonghe, 1997). In case of alumina (Harmer et al., 1979) for example, the activation energy for densification is greater than that for grain growth, and high sintering temperatures the most suitable (Harmer & Brook, 1981).

Kim and Kim (Kim & Kim, 1993) studying the effect of heating rate on shrinkage of pores in yttria-zirconia doped, stated that growth of pores is also inhibited by the fast firing process, helping thus to increase the densification. Searcy (Beruto et al., 1989; Searcy, 1987) suggested that the beneficial effects attributed to the fast firing may be due in part to temperature gradients developed in the sample during heating.

Rate-controlled sintering is more efficient for non-agglomerated powders, in which the microstructure develops relatively homogeneous. However, benefits of these techniques have proved less effective for agglomerated systems. The difficulty of obtaining homogeneous green microstructures using ultra-fine powders, owing to their high degree of agglomeration leads to inhomogeneous, low densification rate and limited final density (Rosen & Bowen, 1988; Inada et al. 1990; Dynys & Hallonen, 1984).

Recently, the availability of many different production routes for ultrafine and nanosized ceramic powders have led research to focus increasingly on the processing of these types of powders. Transformation processes that occur at low temperatures have been observed and studied particularly before or at the beginning of the densification stages of sintering. These processes, which have been reported for coarsening and particle repacking (Chen & Chen, 1996, 1997), affect the subsequent sintering stages. When these processes are controlled, it is possible to obtain dense and fine microstructures. One way to control these processes is optimizing the material's heating curve by pre-treating it at low temperatures.

De Jonghe et al. (Chu et al., 1991; Lin & DeJonghe, 1997a, 1997b) found that pre-heat treatment (50 to 100 hours) at low temperatures (800 °C), in which little or no densification occurs, can improve densification and microstructure of a high purity alumina with and without MgO addition. A consequence of these pre-treatments was reducing the densification rate in the initial stages of sintering. However, benefits of evolving a more homogeneous microstructure are evidenced in the final stages of sintering, allowing a final microstructure refinement. According to DeJonghe et al. (Chu et al., 1991; Lin & DeJonghe, 1997a, 1997b), the pre-treatment leads to more compaction due to the strong increase in the neck formation among particles, promotes the elimination of fine particles, probably through the ripening process of Ostvald and produces a narrower distribution in pore size. These factors make decrease the density fluctuation during sintering, thus favoring the achievement of more uniform microstructures. The best microstructural homogeneity, both in relation to pores and particles, retards the closing of the pore network, so that pores remain open until higher densities inhibiting grain growth more effectively (Lin & DeJonghe, 1997b).

Kim and Kishi (Kim & Kishi, 1996) observed the effect of pre-treatments on strength and subcritical crack growth in alumina. Alumina sintered by hot pressing had 400-500MPa resistance while alumina subjected to a pre-treatment (1000 to 1200 °C for 10 hours) increased their resistance to 750MPa. They concluded that the fracture toughness of grain boundary is increased with pre-treatment and toughness of grain boundary reduces the rate of subcritical crack growth sintering resulting in increased strength of the material. Sato and Carry (Sato & Carry, 1995) studied the effect of particle size and pre-treatment on ultra-fine alumina and found that the pre-treatment delays the start of abnormal grain growth, creating a more uniform microstructure before densification. Chinelatto et al. (Chinelatto et al., 2008) studied the influence of heating curve on the sintering of alumina subjected to high-energy milling and observed that the isothermal treatments at a temperature below the beginning of linear shrinkage cause the fine particles to disappear, narrowing the final grain size distribution.

A new sintering process in two steps was proposed by Chen (Chen & Wang, 2000). The author showed the possibility of obtaining fully dense bodies and sizes of nanosized grains in sintering without applying pressure. This rapid sintering technique inhibits grain growth that occurs in the final stages of sintering and consists of a heating curve in which the ceramic body is subjected to a rapid peak in temperature followed by cooling to the sintering level. Thus, there is densification of the material without the characteristic grain growth. Suppression of grain growth in the final stage of sintering was achieved by exploiting the difference between the kinetics of diffusion in the grain boundary and controlled grain boundary migration rate. Chen and colleagues used the technique of two-step sintering nanosized powders of Y_2O_3 (Wang et al., 2006a), $BaTiO_3$ ferrites and Ni-Cu-Zn (Wang et al., 2006b). Other studies are reported in the literature using the two-step sintering to post nanometric TiO_2 (Mazaheri, 2008a), yttria stabilized zirconia (Mazaheri, 2008b), zirconia (Tartaj, 2009), abrasive alumina with additions of $MgO-CaO-SiO_2$ (Li et al., 2008), alumina-zirconia (Wang et. al., 2009) among others.

According to Chen and Wang (Chen & Wang, 2000; Wang et al., 2006a), in a temperature range called kinetic window, occurs by grain boundary or volumetric diffusion while the grain boundary movement is restricted, so that the densification occurs without, however, growth occurs grain. The sintering temperature in this region results in elimination of residual porosity without grain growth at final stage of work. Suppression of grain growth but not densification is consistent with a network of grain boundaries anchored by triple junction points, which have higher activation energy for migration than grain boundaries (Wang et al., 2006a).

The choice of temperature for both steps is essential for successful sintering. If densities greater than a critical value are reached in the first heating stage, the density of triple junctions decreases, so the effect of the triple points drag mechanism is reduced and the grain growth control is injured in the final sintering stages. On the other hand, if densities are lower than certain critical value, it is not possible to achieve material's densification in the second sintering stage (Chen & Wang, 2000; Hesabi et al., 2009).

Zhou et al. (Zhou et al., 2003) showed that triple junction at large grain sizes is not significant since its volume fraction is negligible compared with the total interface fraction. It is believed that occurs when the passage to the second stage of sintering, the energy in the triple junction during the whole period of time, remains constant. If there is increase in temperature it may be due to increased energy of the system, so there may be a greater mobility of the triple junction in comparison with the grain boundary, so the contour can move freely without any difficulty, indicating a common growth grain. At low temperature, the triple junctions make difficult the movement of grain boundaries not allowing grain growth occurrence (Czubayko et al., 1998).

Nanometric and sub-micrometric alumina powders (Hesabi et al., 2009; Li & Ye, 2006) were also sintered in two steps. Ye and Li (Li & Ye, 2006) found that it is necessary that nanosized alumina powders reach 85% theoretical density in the first stage of sintering, so it can be fully densified at the second level, while Bodisova (Bodisova et al., 2007) showed that density should not be less than 92% theoretical density to achieve full densification without grain growth in the second level for post sub-micron alumina.

2.2 Addition of particles of a second phase

A strategy used to achieve nanometric grain sizes is through addition of solutes or particles of a second phase in single-phase ceramics, which reduce the grain boundary mobility or fix the grain boundary, respectively (Novkov, 2006). This strategy has been used successfully by many researchers. Chaim et al. (Chaim et al., 1998) added 4 wt% trivalent oxides (Y, La, Bi) and tetravalent oxides (Ce, Th) in nanocrystalline zirconia powder and found that Y_2O_3, CeO_2 and ThO_2 inhibit grain growth during sintering. According to Mayo (Mayo, 1996), Hahn et al. added to a powder Y_2O_3 nanocrystalline TiO_2 to limit grain growth. Part of Y_2O_3 dissolved in the regions of grain boundaries and partly reacted with TiO_2 to form a second phase in grain boundaries. These two effects have limited the growth of grains so that the Y_2O_3 sintered without applying pressure reached 90% density with 50nm grain size and Y_2O_3; when adding TiO_2, sintered under the same conditions, it showed 30nm grain size with 99% density.

Recent studies have shown that grain growth inhibition during sintering, which favors increase in mechanical properties of the nanocomposite, occurs by adding small amounts of nanosized zirconia inclusions in a ceramic body of alumina matrix. Grain growth inhibition has also been observed with nanometric inclusions of silicon carbide. However, densification during sintering is difficult by the presence of zirconia in alumina. Other problems were reported in the literature: tendency to particles agglomeration and difficulty to dispersion of nanosized particles of zirconia in alumina matrix, particularly for mechanical mixing methods (Sakka & Hiraga, 1999; Susuki, 2001).

Trombini et al. (Trombini et al., 2007) dispersed powder of alumina and zirconia separately, which allowed them to obtain a complete and homogeneous dispersion of nanosized particles of zirconia in alumina matrix. The Spark Plasma Sintering (SPS) could be used to obtain samples with densities close to theoretical density with very homogeneous microstructure and grain size similar to the initial particle size of powder with at 1300 °C sintering temperature. Pierri et al. (Pierri et al., 2005) observed that the presence of small amounts of zirconia (1 vol%) was sufficient to cause an grain growth inhibition of alumina, allowing the sintering process without application of pressure that results in higher final densities and increased mechanical strength and wear resistance.

3. Experimental procedure

Initially, the alumina powder was processed to remove the hard agglomerates. The following procedure was used: powder was dispersed in isopropyl alcohol with 0.2 w% of PABA (4-aminobenzoic acid) and 0.5 w% of oleic acid. The suspension was submitted to a ball mill during 10 h, using zirconia balls (ball/powder in mass ratio of 2:1) in a polypropylene vial. Suspension was dried at 75°Cand then pulverized and sieved.

For dispersion of zirconia nanometric powder in the alumina powder a ZrO_2 suspension was prepared through traditional balls milling (ZrO_2 balls with 5mm diameter) using 0.5 wt% of deflocculant PABA (4-aminobenzoic acid) in alcoholic medium with a balls/powder mass ratio of 4:1. After 12 hours milling, suspension was separated through the milling and reserved. Simultaneously, Al_2O_3 suspension in alcoholic medium was prepared with 0.2 wt% PABA with a balls/powder ratio of 5:1 for 1 hour in balls mill. 5 vol% ZrO_2 previously prepared were added to this suspension under agitation. Then, final suspension was mixed in conventional balls mill for 22 hours. Finally, 0.5w% oleic acid was added to the suspension and mixed for 2 more hours. The obtained mixtures were dried at room temperature under flowing air.

Prior to sintering experiments, samples of pure alumina were uniaxially pressed under 80 MPa into cylindrical compacts (ø=10 mm, and height of about 5 mm) and isostatically cold-pressed under 200 MPa. Samples were heat-treated at 600°C in air for 1 h to eliminate organic materials. Green density of samples was about 59% of the theoretical density (%TD).

4. Results and discussion

Conventional sintering experiments were carried out at temperature between 900 and 1500°C in air with 2 h dwell time. This sample also was sintered in a Netzsch – DIL 402C dilatometer at 15°C/min constant heating rate in air atmosphere. Based on these results, steps for the sintering were defined. The sintering process was performed in electric furnace (Model Lindberg) in the presence of $MoSi_2$ heating elements in air atmosphere.

In addition to thermal analysis by dilatometry, sintered samples were further characterized by the apparent density taken the Archimedes method as reference, grain size measurements using an image analysis program, and the microstructure was analyzed by scanning electronic microscopy (SEM).

Figure 1 shows the linear shrinkage rate versus temperature during heating in dilatometer at15°C/min constant heating rate. Sintering shrinkage started at 1030°C and maximum shrinkage rate occurred at 1345°C. In figure 1, two different areas can be defined: the first area beginning between 900°C and 1000°C and until approximately 1080°C and refers to the temperature range before sample shrinkage beginning. As shrinkage is directly related to the densification of ceramic body during the sintering process, sintered samples did not begin the densification at temperatures lower than 1080°C having a rearrangement process, coarsening of the particles and appearance of contact points among particles. The second area can be defined as the one where shrinkage occurs, from approximately 1080°C to 1500°C. In this area, shrinkage rate reaches the maximum value at approximately 1350°C.

The sintering temperature effect on the densification and grain growth of compacts sintered at temperature ranging between 900 and 1500°C for 1 hour is shown in figure 2. No significant densification was observed below 1030°C confirming the dilatometric results (figure 1). Densification was accelerated at the temperature between 1100°C and 1350°C, without, however, presenting great grain growth. At higher temperature, densification was minimal but the grain growth was fast. Final grain size of the nearly fully dense structure was higher than 1800 nm. While relative density increased from 95% to 99.2% with increase in temperature from 1300°C to 1500°C, the average grain size became coarser from 480 nm to 1800 nm; in other words, there was more than 250% increase in grain size.

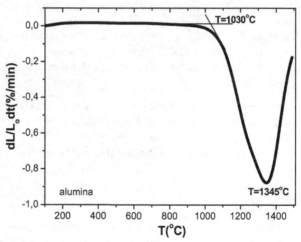

Fig. 1. Linear shrinkage rate versus temperature during heating in dilatometer at 15°C/min constant heating rate.

It has reported that dispersed open pores can pin grain boundaries and hinder grain-boundary migration in the second stage of sintering, for which the grain growth is suppressed (German, 1996). In contrast, a very sharp ascending of grain size is observable in the final sintering stage (relative density above 90% TD); however, there is remarkable increase in density. It has been confirmed that open pores referring to the intermediate stage of sintering collapse to form the closed ones after the final stage starts. Such a collapse results in a substantial decrease in pore pinning, which triggers the accelerated grain growth.

Considering the results of sintering experiments (figures 1 and 2) and to suppress the accelerated grain growth at the final sintering stage, two different sintering heating curves were applied to produce densification of Al_2O_3 compacts. These experiments were carried out using 15°C/min heating rate.

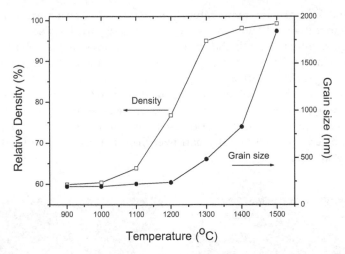

Fig. 2. Density and grain size of Al$_2$O$_3$ compacts after sintering at various temperatures for 1 hour.

In the first sintering heating curve one hypothesis was assumed: the maximization of final density with minimum grain growth could be achieved by improving the narrowing of grain size distribution at a pre-densification sintering stage and producing the final densification at a maximum densification rate. To confirm this hypothesis, a temperature below the onset of the densification process was chosen. This effect can be observed in samples with the first step at 1050°C. Samples produced by the first step at 1050°C followed by a second step at 1500°C showed significantly smaller final grain sizes as shown in figure 3 (Chinelatto et al., 2010).

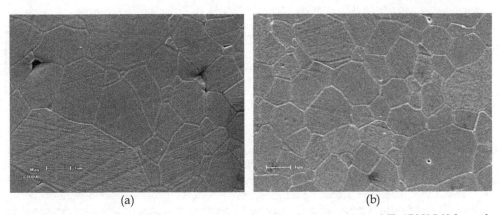

(a)   (b)

Fig. 3. SEM micrographs of alumina samples after two-step sintering: a) T=1500°C/2 h; and b) T=1050°C/2 h and T=1500°C/2 h.

Figure 4 shows micrographs of surface fracture of alumina compacts, one heated at 1050°C and cooled immediately upon reaching that temperature; and the other heated to the same temperature and kept at such temperature for 2 hours. The heat treatment made finest particles to disappear and slightly coarsened the coarsest particles, decreasing the specific surface area and slightly increasing the mean grain size as indicated in Table 1. De Jonghe et al. (Lin & DeJonghe, 1997a, 1997b) suggested that during the first step, coarsening of the microstructure by surface diffusion, vapor transport, or some combination of these mechanisms produces a more uniform microstructure by an Ostwald ripening process. The evolution to a more homogeneous microstructure can be expected from the trend of porous system to evolve towards a quasi-steady state structure. Such steady-state structural distributions are generally significantly narrower than that usually produced in a powder compact (Chu et al., 1991).

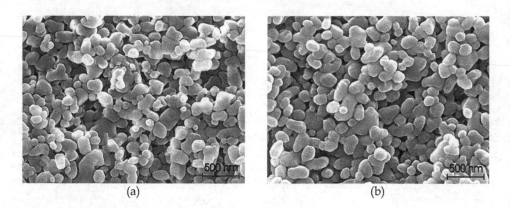

(a)   (b)

Fig. 4. SEM fracture surfaces of Al$_2$O$_3$ compacts: (a) heated at 1050°C and cooled immediately e (b) heated at 1050°C for 2 hours.

	1050°C	1050°C/2 hours
Superficial area (m²/g)	12.3	10.5
Mean grain size (nm)	119 ± 33	133 ± 28

Table 1. **Superficial area and mean grain size of particles.**

Other heating curves were developed applying sintering curves coherent with the temperature ranges in which the two processes, i.e., narrowing grain size distribution and final densification, were expected to occur. The following conditions were defined for the sintering heating curves: the first step for alumina was at 1050°C and 1000°C and the second step was at the maximum sintering temperature of 1350°C. Table 2 describes the sintering conditions and findings regarding density and average grain size of samples produced in the two-step sintering experiments.

Changes in the relative density and mean grain size with the holding time obtained are shown in Figure 5. Increased holding time results in increase of relative density and decrease in mean grain size.

Sintering procedure	Relative Density (%TD)	Mean Grain Size (nm)
TSS1 - T_1=1000°C/3h and T_2=1350°C/3h	93.8	797.4
TSS2 - T_1=1000°C/6h and T_2=1350°C/3h	94.2	763.5
TSS3 - T_1=1000°C/9h and T_2=1350°C/3h	94.6	683.5
TSS4 - T_1=1050°C/3h and T_2=1350°C/3h	93.9	717.1
TSS5- T_1=1050°C/6h and T_2=1350°C/3h	94.1	685.1
TSS6 - T_1=1050°C/9h and T_2=1350°C/3h	94.2	659.3

Table 2. **Sintering procedure and results of relative density (%TD) and mean grain size of alumina samples.**

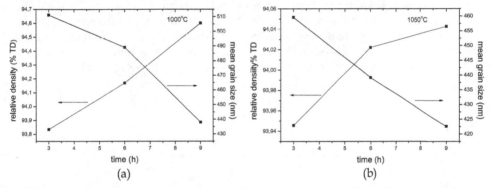

Fig. 5. Relative density and mean grain size of alumina compacts sintered versus holding time at: (a) 1000°C and (b) 1050°C.

According to Lin and Dejonghe (Lin & DeJonghe, 1997a, 1997b), with the steps at low temperature, the onset of densification is delayed due to the elimination of the finest particles (and smallest pores associated with them) during the first step. The local densification associated with the finest particles in the conventional sintering is significantly reduced in compacts subjected to the first step. Thus, removal of the finest particles due to the first step reduces the differential densification and formation of densest regions in the early sintering stages. This fact causes reduction in density fluctuations in the compact and promotes a more homogeneous final microstructure.

The other two-step sintering is based on works of Chen and Wang (Chen &Wang, 2000), in which samples are first heated to a higher temperature to achieve intermediate density, and then cooled down and kept at lower temperature until they are dense. A pre-requisite for successful densification during the second step of sintering is that pores become subcritical and unstable against shrinkage.

Chen and Wang (Chen &Wang, 2000) have explained that to achieve densification without grain growth, grain-boundary diffusion needs to remain active, while the grain-boundary migration must to be suppressed. A mechanism to inhibit grain-boundary movement is a triple-point (junction) drag. Consequently, to prevent accelerated grain growth, it is essential to decrease grain-boundary mobility. The grain growth entails a competition between grain-boundary mobility and junction mobility. Once the latter becomes less at low temperatures in which junctions are rather motionless, the mentioned drag would occur. Therefore, the grain growth is prohibited. Network mobility follows the grain-boundary mobility at high temperatures. At low temperatures, junction mobility dominates. Below the temperature at which the two rates become equal, junction mobility is essentially reduced despite the considerable grain-boundary diffusion.

Figure 2 shows that grain growth is most intense at temperatures above 1400°C. Since samples conditions after the first stage affect the second stage of sintering, grain growth resulting from heating in the first stage must be avoided. Thus, the temperature chosen for the first stage of sintering was 1400 °C.

Figure 6 shows the behavior of relative density (%TD) versus temperature for sintering at constant heating rate of alumina. Density of alumina when temperature reaches 1400°C is 81% TD. The relative density during sintering was determined from green density (dv) and measured shrinkage ($\Delta L/Lo$), using the approximate Eq. (1), assuming that deformation is isotropic and all axial strain is devoted to specimen's densification (Ray, 1985).

$$d_i = \frac{d_v}{(1+\frac{\Delta L}{L_o})^3} \tag{1}$$

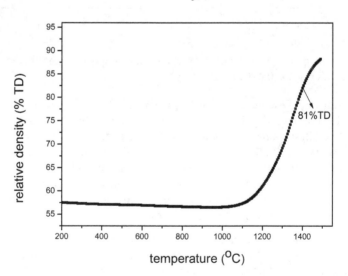

Fig. 6. Variation of relative density (% TD) versus temperature for alumina sintered at 15°C/min until the temperature of 1500 °C.

SEM micrograph of alumina when it reaches 1400°C in the first step of sintering is showed in Figure 7. The mean grain size of alumina in this condition is about 330 nm.

Fig. 7. SEM micrograph of alumina sintered at 1400°C.

To choose the temperature for the second step T2, it is necessary to choose a temperature in which volume diffusion or grain boundary diffusion operate while the grain boundary movement is restricted (Mazaheri et al., 2008). Therefore, the second step temperatures were 1260°C and 1300°C. Sintering conditions and results of density relative and mean grain size are presented in table 3.

Sintering procedure	Relative density (%TD)	Mean grain size (nm)
TSS7 - T_1=1400°C and T_2=1260°C/3h	91.0	518.8
TSS8 - T_1=1400°C and T_2= 1260°C/6h	92.1	579.2
TSS9 - T_1=1400°C and T_2= 1260°C/9h	93.1	647.8
TSS10 - T_1=1400°C and T_2= 1300°C/3h	95.9	668.5
TSS11 - T_1=1400°C and T_2= 1300°C/6h	96.5	692.5
TSS-12 - T_1=1400°C and T_2= 1300°C/9h	96.6	718.7

Table 3. Sintering procedure and results of relative density (%TD) and mean grain size of alumina samples.

According to the results of the second step of TSS10, TSS11 and TSS12, holding the samples at 1300°C resulted in accentuate densification. Diffusive mechanisms that seem to be time dependent are therefore active at this stage. Grain-boundary diffusion and volumetric diffusion are possibly responsible for the shrinkage of the samples. On the other hand, TSS7, TSS8 and TSS9 do not lead to a dense structure, showing the inactivity of the grain-boundary diffusion at 1260°C. Considering all these facts, one can infer that 1300°C is the minimum temperature after which the grain boundary diffusion mechanism dominates.

Due to the relatively low temperature of the second stage (1260°C), densification stops before reaching a fully dense sample. A similar trend has also been reported for the two-step sintering behavior of Y_2O_3 (Wang and Chen, 2006) and ZnO (Mazaheri et. al., 2008) confirming that the reason for exhaustion in the second stage of densification is attributed to low temperature which retards grain-boundary diffusion as the sintering mechanism.

It can be seen in figure 8 that density variation results in increased grain size of the sample, showing that this condition is not yet the ideal to control the grain size in two-step sintering.

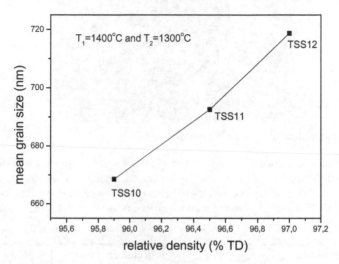

Fig. 8. Grain size/relative density trajectory obtained by two-step sintering T_1=1400°C and T_2=1300°C.

On the other hand, comparing the two-step sintering with conventional sintering, it is observed that the two-step sintering is efficient to control the grain growth. Figure 9 shows the micrographs of alumina sintered at 1500°C for 2 hours and sintered at TSS3 and TSS12 conditions.

The heating curve control, through using steps of sintering, associated with control of grain size by addition of nanometric zirconia inclusions is also control the microstrucuture in conventional sintering.

Figure 10 shows the linear shrinkage rate as function of temperature for alumina-5%vol zirconia at 15°C/min constant heating rate and 1500°C. The presence of zirconia particles increases the maximum densification rate temperature; for alumina this temperature is 1350°C (figure 1) and for alumina-zirconia the temperature is increased for 1440°C. The temperature at the beginning of shrinkage process is also altered from 1030 to 1210°C with the addition of zirconia particles. Zirconia inclusions hinder the movement of grain boundary, reducing the densification rate and grain growth (Hori et al., 1985; Liu et al., 1998; Stearns & Harmer, 1996). In the figure 11 (a) and (b), that shows the micrographs of samples of alumina-zirconia and alumina, respectively, sintered at 1500°C for 2h, the influence of zirconia on microstructure evolution is noted through observing the grain growth behavior. The addition of nanometric zirconia is very efficient to promote a controlled grain growth. The inhibitive trend is due to the pinning effect which is associated with locations of small zirconia particles at grain boundaries or triple junctions of alumina.

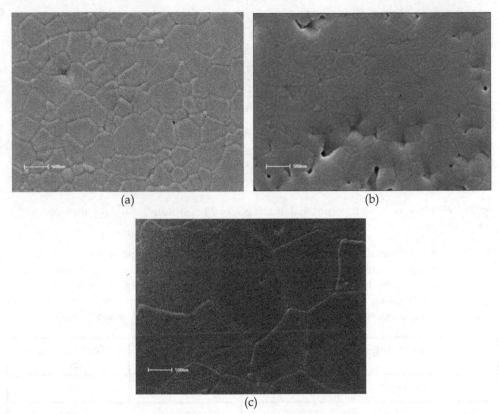

(a)   (b)

(c)

Fig. 9. SEM micrographs (a) TSS3; (b) TSS12 and (c) CS.

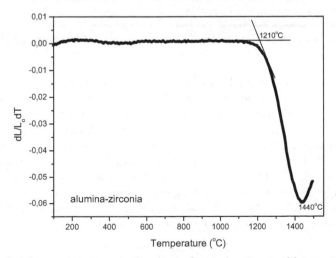

Fig. 10. Linear shrinkage rate versus temperature during heating in dilatometer at 15°C/min constant heating rate for alumina-zirconia compacts.

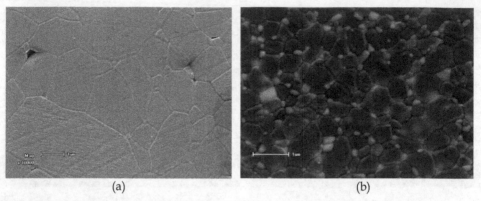

(a)   (b)

Fig. 11. SEM images of sample sintered at 1500°C for 2 h: (a) alumina and (b) alumina-zirconia.

The heating curve control, combined with the presence of nanoparticles inclusions can further optimize the microstructure control. Table 4 shows the sintering procedure and results of relative density (%TD) and mean grain size for alumina-zirconia samples. Results for TSS13 and TSS14 conditions show that the two-step sintering promoted reduction of the mean grain size compared to the conventional sintering (CS1) (Manosso et al., 2010).

Sintering procedure	Relative density (%TD)	Mean grain size (nm)
CS – T=1500°C/2h	99.0	550
TSS13 - T_1=1460°C/h and T_2=1350°C/3h	97.8	330
TSS14 - T_1=1300°C/2h and T_2=1460°C/2h	99.7	410

Table 4. Sintering procedure and results of relative density (%TD) and mean grain size of alumina-zirconia samples.

The microstructure of the sample heated at 1460°C and cooled immediately and the sample sintered under TSS13 conditions are showed in Figure 12. It can be noticed an initial densification and 77%DT relative density for this sample. When the sample was heated at 1460°C and cooled down to 1350°C (TSS13 condition) the sample could be densified without grain growth (see table 2). It means that, 77% DT reached density in the first step at high temperature for this sample can be considered the critical density. In spite of the smaller grain size presented by TSS1 condition, its relative density was lower than densities of TSS14 and CS conditions. It suggests that the time of soaking in the second step can still be prolonged. Many studies (Tarjat & Trajat, 2009; Mazaheri et al., 2008) have been demonstrated that long times in the second steps allowed the total densification without grain growth.

In the TSS14 condition, pre-densification sintering stage at 1300°C for 2 hours was effective in grain growth control and final densification. Figure 13 (a), (b) and (c) presents the microstructure of alumina-zirconia sintered under CS1, TSS13 and TSS14 conditions, respectively. These micrographs confirmed that the two-step sintering used have been efficient to the sintering process control. It can be observed that the sample conventionally sintered presents larger grain size. Finally, it was observed that the step of sintering with addition of inclusions is also efficient in grain growth.

(a)   (b)

Fig. 12. SEM micrographs of sintered samples: (a) T_1=1460°C and cooled and (b) TSS13 condition.

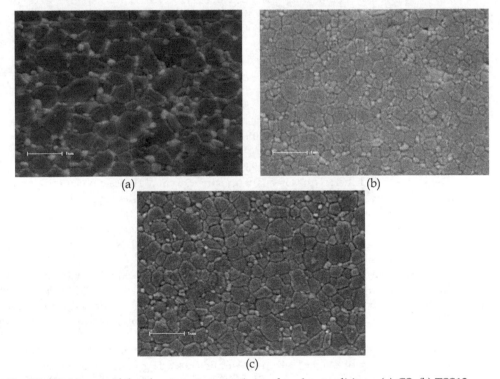

(a)   (b)

(c)

Fig. 13. SEM image of the alumina-zirconia sintered under conditions: (a) CS; (b) TSS13; (c) TSS14.

5. Conclusion

The introduction of isothermal treatments in the heating curve at temperatures below the beginning of an accentuated shrinkage process influenced the development of the final microstructure, promoting a microstructural refinement of particles compacts.

The heating curve control is a simple and efficient method to control ceramic microstructure, although it is difficult to achieve an optimum condition for accessing a successful regime. The main characteristics of the heating curve control are: nanostructured ceramics can be obtained with nearly full densities; it is not necessary sophisticated and unavailable equipment, like those used for spark plasma sinterig and hot isostatic pressing; and and it is possible to achieve fully dense structures at lower temperatures.

The heating curve control, combined with the presence of nanoparticles inclusions can further optimize the microstructure control. Fine grains in the sintering induce a pinning effect on grain boundary migration and the degree of grain growth during sintering is effectively reduced.

6. References

Allen, A. J.; Krueger, S.; Skandan, G.; Long, G. L.; Hahn, H.; Kerch, H. M.; Parker, J. C. & Ali, M. N. (1996). Microstructural Evolution during the Sintering of Nanostructured Ceramic Oxides *J. Am. Ceram. Soc.*, Vol. 79, No. 5, (May 1996), pp. (1201-1212), ISSN 0002-7820.

Averback, R. S.; Höfler, H. J.; Hahn, H. & Logas, J. C. (1992). Sintering and Grain Growth in Nanocrystalline Ceramics, *Nano Mat.*, Vol. 1, No. 1, (March-April 1992) ,pp. (173-178), ISSN 0965-9773.

Bernard-Granger, G. & Guizard, C. (2007). Spark Plasma Sintering of a Commercially Available Granulated Zirconia Powder: I. Sintering Path and Hypotheses about the Mechanism(s) Controlling Densification, *Act. Mater.*, Vol. 55, No. 10, (June 2007), pp. (3493-3504), ISSN 1359-6454.

Beruto, D.; Botter, R. & Searcy, A. W. (1989). The Influence of Thermal Clycling on Densification Further: Tests of a Theory, In: Ceramic Transaction, v.1 Ceramic Powder Science IIB. Ed by Fuller, E. R. Jr.; Husner, H & Messing, G.L., pp. (911-918), Am. Ceram. Soc. Inc., ISBN: 0916094316 Westervillw, OH.

Bodisova, K.; Sajgalik, P.; Galusek, D. & Svancare, P. (2007). Two-Stage Sintering of Alumina with Submicrometer Grain Size, *J. Am. Ceram. Soc.*, Vol. 90, No. 1, (January 2007), pp. (330-332), ISSN 0002-7820.

Brook, R.J. (1982). Fabrication Principles for the Production of Ceramics with Superior Mechanical Properties, *Proc. Br. Ceram. Soc*, Vol. 32, pp. (7-24), ISSN 0524-5141.

Chaim, R.; Basat, G. & Kats-Demyanets, A. (1998). Effect of Oxide Additives on Grain Growth During Sintering of Nanocrystalline Zirconia Alloys, *Mater. Let.*, Vol. 35, No. 3-4 , (May 1998), pp. (245-250), ISSN 0167-577X.

Chakravarty, D.; Bysakh, S.; Muraleedharan, K.; Rao, T. N. & Sundaresan, R. (2008). Spark Plasma Sintering of Magnesia-Doped Alumina with High Hardness and Fracture Toughness, *J. Am. Ceram. Soc.*, Vol. 91, No. 1, (January 2008), pp. 203-208 ISSN 0002-7820.

Chen, P. L. & Chen, I. W. (1996). Sintering of Fine Oxide Powders: I, Microstructural Evolution, *J. Am. Ceram. Soc.*, Vol. 79, No. 12, (December 1996), pp. (3129-3141), ISSN 0002-7820.

Chen, P. L. & Chen, I. W. (1997). Sintering of Fine Oxide Powders: II, Sintering Mechanism, *J. Am. Ceram. Soc.*, Vol. 80, No. 3, (March 1997), pp. (637-645), ISSN 0002-7820.

Chen, I. W. & Wang, X. H. (2000). Sintering dense nanocrystalline ceramics without final-stage grain growth, *Nature*, Vol. 404, (March 2000), pp. (168-171), ISSN 0028-0826.

Chinelatto, A. S. A.; Manosso, M. K.; Pallone, E. M. J. A.; Souza, A. M. & Chinelatto, A.L. (2010). Effect of the Two-Step Sintering in the Microstructure of Ultrafine Alumina, *Adv. Sci.Tech.*, Vol. 62, (October 2010), pp. (221-226), ISSN 1662-0356.

Chinelatto,A. S. A.; Pallone, E. M. J. A.; Trombini, V. & Tomasi, R. (2008). Influence of Heating Curve on the Sintering of Alumina Subjected to High-Energy Milling. *Ceram. Int.* Vol. 34, No. 8, (December 2008), pp. (2121–2127), ISSN 0272-8842.

Chu, M.Y.; DeJonghe, L.C.; Lin, M.K.F. & Lin, F.J.T. (1991). Precoarsing to Improve Microstructure and Sintering of Powder Compacts, *J. Am. Ceram. Soc.*, Vol. 74, No 11, (November 1991), pp. (2902-2911), ISSN 0002-7820.

Czubayko, U.; Sursaeva, V.G.; Gottstein, G. & Shvindlerman, L. S. (1998). Influence of Triple Junctions on Grain Boundary Motion, *Act. Mater.* Vol. 46, No. 16, (October 1998), pp. (5863-5871), ISSN 1359-6454.

Dynys, F. W. & Hallonen, T.W. (1984). Influence of Aggregates on Sintering, *J. Am. Ceram. Soc.*, Vol. 67, No. 9, (September 1984), pp. (596-601), ISSN 0002-7820.

Erkalfa, H.; Misirk, Z. & Baykara, T. (1996). Effect of Additives on the Densification and Microstructural Development of Low-Grade Alumina Powders, *J. Mat. Proc. Tec.*, Vol. 62, No. 1-3, (November 1996), pp.(108-115), INSS 0924-0136.

Gao, L.; Hong, J. S.; Miyamoto, H & Torre, D. D. L.(2000). Bending Strength and Microstructure of Al_2O_3 Ceramics Densified by Spark Plasma Sintering, *J. Eur. Ceram. Soc.*, Vol. 20, No. 12, (November 2000), pp. (2149-2152) ISSN 0955-2219.

German, R. M. (1996). *Sintering: Theory and Practice.* Ed. John Wiley & Sons, ISBN 978-0471057864, New York.

Greer, A. L. (1998). Nanostrucuture Materials – from Fundamentals to Applications. *Mat. Sci. Forum*, Vol. 269-272, pp. (3 -10), ISSN 0255-5476.

Hahn, H. (1993). Microstrucutre and Properties of Nanostructured Oxides, *Nano. Mat.*, Vol. 2, No. 3, (May-June 1993), pp. (251-265), ISSN 0965-9773.

Hahn, H.; Logas, J & Averback, R. S. (1990). Sintering Characteristics of Nanocrytalline TiO_2, *J. Mater. Res.*, Vol. 5, No. 3, (May 1990), pp. (609-614), 1990, ISSN 0884-2914.

Harmer, M.P. & Brook, R.J. (1981). Fast Firing Microstructural Benefits, *J. Br. Ceram. Soc.*, Vol. 80, No. 5, pp. (147-48), ISSN 0307-7357.

Harmer, M. P.; Roberts, E. W. & Brook, R. J. (1979). Rapid Sintering of Pure and Doped α-Al_2O_3, *J. Br. Ceram. Soc.*, Vol. 78, No. 1, pp. (22-25), ISSN 0307-7357.

He, Z. & Ma, J. (2000). Grain Growth Rate Constant of Hot-Pressed Alumina Ceramics, *Materials Letters*, Vol. 44, No.1, (May 2000), pp.14-18, ISSN 0167-577X.

Hesabi, Z. R.; Haghighatzadeh, M. ; Mazaheri, M. ; Galusek, D. & Sadrnezhaad, S. K. (2009). Suppression of Grain Growth in Sub-Micrometer Alumina via Two-Step Sintering Method, *J. Eur. Ceram. Soc.*, Vol. 29, No. 8, (May 2009), pp. (1371-1377), ISSN 0955-2219.

Hori, S.; Kurita, R.; Yoshimura, M. & Somiya, S. (1985). Suppressed grain growth in final-stage sintering of Al2O3 with dispersed ZrO2 particles, *J. Mat. Sci. Let.*, Vol. 4, No. 9, (September 1985), pp. (1067-1070), ISSN 0261-8028.

Inada, S.; Kimura, T. & Yamaguchi, T. (1990). Effect of Green Compact Structure on the Sintering of Alumina, *Ceram. Int.*, Vol. 16, No. 6, pp. (369-373), ISSN 0272-8842.

Inoue, A. & Masumoto, T. (1993). Nanocrystalline Alloys Produced by Crystallization of Amorphous Alloys, In: *Current Topics in Amorphous Materials: Physics an Technology*, Ed. by Y.Sakurai, Y.Hamakawa, Y.Masumoto, K.Shirae and K. Suzuki, pp. (177-184), Elsevier Science Ltd, ISBN 9780-444815767, Amsterdam.

Kim, B. N. & Kiski, T. (1996). Strengthening Mechanism of Alumina Ceramics Prepared by Precoarsening Treatments, *Mat. Sci. Eng. A*, Vol. 215, No. 1-2, (September 1996), pp. (18-25). ISSN 0921-5093.

Li, J. & Ye, Y. (2006). Densification And Grain Growth of Al2O3 Nanoceramics During Pressureless Sintering, *J. Am. Ceram. Soc.*, Vol. 89, No. 1, (January 2006), pp. (139-143), ISSN 0002-7820.

Li, Z.; Li, Z; Zhang, A. & Zhu, Y. (2008). Two-Step Sintering Behavior of Sol-Gel Derived Nanocrystalline Corundum Abrasive with $MgO-CaO-SiO_2$ Additions, *J. Sol. Gel Sci. Technol.*, Vol. 48, No. 3, (December 2008), pp. (283-288), ISSN 0928-0707.

Liao, S. C.; Chen, Y, J; Kear, B. H. & Mayo, W. E. (1998). High Pressure/Low Temperature Sintering of Nanocrystalline Alumina, *Nano Mater.*, Vol. 10, No. 6, (August 1998), pp. (1063-1079), ISSN 0965-9773.

Lim, L. C.; Wong, P. M. & Ma, J. (1997). Colloidal Processing of Sub-Micron Alumina Powder Compacts, *J. Mat. Proc. Tec.*, Vol. 67, No. 1-3, (May 1997), pp. (137–142), INSS 0924-0136.

Lin, F. J. T. & DeJonghe, L. C. (1997a). Initial Coarsening and Microstructural Evolution of Fast-Fired and MgO-Doped Alumina, *J. Am. Ceram. Soc.*, Vol. 80, No. 11, (November 1997), pp. (2891-2896), ISSN 0002-7820.

Lin, F. J. T. & DeJonghe, L. C. (1997b). Microstructure Refinement of Sintered Alumina by Two-Step Sintering Technique, *J. Am. Ceram. Soc.*, Vol. 80, No. 10, (October 1997), pp. (2269-2277), ISSN 0002-7820.

Liu, G. L.; Qiu, H.; Todd, R.; Brook, R. J. & Guo, J. K. (1998). Processing and Mechanical Behavior of Al2O3/ZrO2 Nanocomposites, *Mat. Res. Bull.*, Vol. 33, No. 2, (February 1998), pp. (281-288), ISSN 0025-5408.

Manosso, M. K.; Pallone, E. M. J. A.; Souza, A. M.; Chinelatto, A.L. & Chinelatto, A. S. A. (2010). Two-Steps Sintering of Alumina-Zirconia Ceramics, *Mat. Sci. For.*, Vol. 660-661, (October 2010), pp. (819-825), ISSN 1662-9752.

Mayo, M. J. (1996). Processing of Nanocrystalline Ceramics from Ultrafine Particles, *Int. Mat. Rev.*, Vol. 41, No. 3, (January 1996), pp. (85-115), ISSN 0950-6608.

Mazaheri, M; Hesabi, Z. R. & Sadrnezhaad, S. K. (2008). Two-Step Sintering of Titania Nanoceramics Assisted by Anatase-to-Rutile Phase Transformation, *Scr. Mat.*, Vol. 59, No. 2, (July 2008), pp. (139-142), ISSN 1359-6462.

Mazaheri, M.; Simchi, A. & Golestani-Fardi, F. (2008). Densification and Grain Growth of Nanocrystalline 3Y-TZP During Two-Steps Sintering, *J. Eur. Ceram. Soc.*, Vol. 28, No. 15, (November 2008), pp. (2933-2939), ISSN 0955-2219.

Mazaheri, M.; Zahedi, A. M. & Sadrnezhaad, S. K. (2008). Two-Step Sintering of Nanocrystalline ZnO Compacts: Effect of Temperature on Densification and Grain Growth, *J. Am. Ceram. Soc.*, Vol. 91, No. 1, (January 2001), pp.(56-63), ISSN 0002-7820.

Mazaheri, M.; Zahedi, A. M..; Haghighatzadeh, M. & Sadrnezhaad, S.K. (2009). Sintering of Titania Nanoceramic: Densification and Grain Growth, *Ceram. Int.*, Vol. 35, No. 2, (March 2009), pp. (685-691), ISSN 0272-8842.

Morris, D. G. (1998).What Have We Learned about Nanoscale Materials? The Past and Future. *Mat. Sci. Forum*, Vol. 268-272, pp. (11-14), ISSN 0255-5476.

Novikov, V. Y. (2006). Grain Growth Controlled by Mobile Particles on Grain Boundaries, *Scr. Mat.*, Vol. 55, No. 3, (August 2006), pp.(243-246), ISSN 1359-6462.

Pierri, J. J.; Maestrelli, S. C.; Pallone, E. M. J. A. & Tomasi, R. (2005). Dispersão de Nanopartículas de ZrO_2 Visando Produção de Nanocompósitos de ZrO_2 em Matriz de Al_2O_3, *Cerâmica*, Vol. 51, No. 317, (Jan-Mar 2005), pp. (08-12), ISSN 0366-6913.

Porat, R.; Berger, S. & Rosen, A. (1996). Dilatometric Study of the Sintering Mechanism of Nanocrystalline Cemented Carbides, *Nano Mater.*, Vol. 7, No. 4, (May-June 1996) pp. (429-436), ISSN 0965-9773.

Ragulya, A.V. & Skorokhod, V.V. (1995). Rate-Controlled Sintering of Ultrafine Nickel Powder, *Nano. Mat.*, Vol. 5, No. 7-8, (September-December 1995), pp. (835-843), ISSN 0965-9773.

Rosen, A. & Bowen, H.K. (1988). Influence of Various Consolidation Techniques on the Green Microstructure and Sintering Behavior of Alumina Powders, *J. Am. Ceram. Soc.*, Vol. 71, No. 11, (November 1988), pp. (970-977), ISSN 0002-7820.

Sakka, Y. & Hiraga, K. (1999). Preparation Methods and Superplastic Properties of Fine-Grained Zirconia and Alumina Based Ceramics, *Nippon Kagaku Kaishi*, Vol. 8, pp. (497–508), ISSN 0369-4577.

Sigmund, W.M.; Bell, N.S. & Bergström, L. (2000). Novel Powder-Processing Methods for Advanced Ceramics, *J. Am. Ceram. Soc.*, Vol. 83, No. 7, (July 2000), pp. 1557–1574, ISSN 0002-7820.

Suzuki, T.; Sakka, Y.; Nakano, K. & Hiraga, K. (2001). Effect of Ultrasonication on the Microstructure and Tensile Elongation of Zirconia-Dispersed Alumina Ceramics Prepared by Colloidal Processing, *J. Am. Ceram. Soc.*, Vol. 84, No. 9, (September 2001), pp.(2132–2134), ISSN 0002-7820.

Sato, E. & Carry, C. (1995). Effect of Powder Granulometry and Pré-Treatment on Sintering Behavior of Submicron-Grained α-Alumina, *J. Eur. Ceram. Soc.*, Vol. 15, No. 1, pp. (9-16), ISSN 0955-2219.

Searcy, A. W. (1987). Theory for Sintering in Temperature Gradients: Role of Long-Range Mass Transport, *J. Am. Ceram. Soc.*, Vol. 70, No. 3, (March 1987), pp. (C61-C62), ISSN 0002-7820.

Stearns, L. & Harmer, M. P. (1996). Particle-Inhibited Grain Growth in Al_2O_3-SiC: I, Experimental Results, *J. Am. Ceram. Soc.*, Vol. 79, No. 12, (November 1996), pp. (3013-3019), ISSN 0002-7820.

Suryanarayana, C. (1995). Nanocrystalline Materials , *Int. Mat. Reviews*, Vol. 40, No. 2, pp. (41-64), ISSN 0950-6608.

Tartaj, J. & Tartaj, P. (2009). Two-Stage Sintering of Nanosize Pure Zirconia, *J. Am. Ceram. Soc.*, Vol. 92, No. Supplement s1, (January 2009), pp. (S103-S106). ISSN 0002-7820.

Trombini, V.; Pallone, E. M. J. A.; Munir, Z. A. & Tomasi, R. (2007). Spark Plasma Sintering (SPS) de Nanocompósitos de Al_2O_3-ZrO_2, *Cerâmica*, Vol. 53, No. 325 , (Jan-Mar 2007), pp. (62-67), ISSN 0366-6913.

Wang, C. J.; Huang, C. Y. & Wu, Y. C. (2009). Two-step Sintering of Fine Alumina–Zirconia Ceramics, *Ceram. Int.*, Vol. 35, No. 4, (May 2009), pp. (1467-1472), ISSN 0272-8842.

Wang, X. H.; Chen, P. L. & Chen, I. W. (2006a). Two-Step Sintering of Ceramics with Constant Grain-Size, I. Y_2O_3, *J. Am. Ceram. Soc.*, Vol. 89, No 2, (February 2006), pp. (431-437), ISSN 0002-7820.

Wang, X. H.; Deng, X. Y.; Bai, H. I.; Zhou, H.; Qu, W. G.; Li, L.T. & Chen, I.W. (2006b). Two-Step Sintering of Ceramics with Constant Grain-Size, II.BaTiO3 and Ni-Cu-Zn Ferrite, *J. Am. Ceram. Soc.*, Vol. 89, No. 2, (February 2006), pp.(438-443), ISSN 0002-7820.

Weibel, A.; Bouchet, R.; Denoyel, R. & Knauth, P. (2007). Hot Pressing of Nanocrystalline TiO_2 (Anatase) Ceramics with Controlled Microstructure, *J. Eur. Ceram. Soc.*, Vol. 27, No. 7, pp. (2641-2646), ISSN 0955-2219.

Zhou, Y.; Erb, U.; Aust, K. T. & Palumbo, G. (2003). The Effects of Triple Junctions and Grain Boundaries on Hardness and Young Modulus in Nanostructured Ni-P, *Scr. Mater.*, Vol.48, No. 6, (March 2003), pp. (825-830), ISSN 1359-6462.

Zhou, Y.; Hirao, K.; Yamauchi, Y. & Kanzaki, S. (2004). Densification and Grain Growth in Pulse Electric Current Sintering of Alumina, *J. Eur. Ceram. Soc.* Vol. 24, No. 12, pp. (3465-3470), ISSN 0955-2219.

3

Ba$_{1-x}$Sr$_X$TiO$_3$ Ceramics Synthesized by an Alternative Solid-State Reaction Route

R.A. Vargas-Ortíz, F.J. Espinoza-Beltrán and J. Muñoz-Saldaña
Centro de Investigación y de Estudios Avanzados del IPN,
Unidad Querétaro, Libramiento Norponiente No. 2000,
Fracc. Real de Juriquilla, CP Querétaro, Qro.
México

1. Introduction

All materials respond to stimulus, whether it be an electric field, mechanical stress, heat or light. The manner and degree to which they respond varies and is often what determines which material is selected for a given application. On the most basic level, elastic materials deform in response to mechanical stress and return to their original form when the load is removed. Other materials conduct electricity in response to an applied voltage. Both of these are well-known phenomena, and materials with such behaviors are sometimes called "trivial". On the other hand are pyroelectric and piezoelectric materials, which generate an electric field with a stimulus of heat or mechanical stress, respectively (unexpected phenomenon) and are called "smart" or "functional" materials. Ferroelectric materials are materials that exhibit piezoelectricity and pyroelectricity, as well as the phenomenon which gives them their name (ferroelectricity).

Due to their unique properties, ferroelectric materials are widely used in all areas of electronics and microelectronics, such as cellular phones, computers, cars, airplanes and satellites [KENJI, BUCHANAN]. They have a high discharge dielectric constant (ε) [SHEPARD, RADHESHYAM ,ZHIN], which allows them to be used in high permitivity dielectric devices. Their pyroelectric behaviour is used in heat sensors [PADMAJA, YOO, WHATMORE], and their piezoelectricity is applied in devices like resonadores, sonars, horns, and actuators [GURURAJA, YAMASHITA 1997, YAMASHITA 1998, CHEN]. A combination of their properites are applied in electro-optical devices such as controlable diffraction grids, waveguides, etc. [HEIHACHI, HAMMER, BLOMQVIST]. They are also used in dynamic random access memory (DRAM) [KINGON 2000, KINGON 2006, KOTECKI] and non-volatile memory (NVRAM) [MASUI, KOHLSTEDT].

Many "novel" materials known today were developed many decades ago. Ferroelectric materials are no exception, having been discovered more than seven decades ago. Valasek reported the first ferroelectric material, Rochelle salt (potassium or sodium tetrahydrate tartatre, $KNaC_4H_4O_6 \bullet 4H_2O$) in 1921 [VALASEK]. Subsequently, potassium dihydrogen phosphate (KH_2PO_4) was identified by Busch and Scherrer in 1935 [BUSCH], and barium titanate ($BaTiO_3$ or BT) was noted for its unusual dielectric properties by Wainer and

Salomon in 1942-43 [WAINER]. The discovery of ferroelectricity in ceramics from the BaO-TiO$_2$ system was extremely important, as it was the first ferroelastic made from simple oxide materials. Since the discovery of BaTiO$_3$, several other oxide-based ferroelectric materials have been developed, such as strontium titanate (SrTiO$_3$ or ST), lead zirconate titanate (PZT), lead titanate (PbTiO$_3$, PT), lithium tantalate (LiTaO$_3$) phosphate, and potassium titanyl (KTiOPO$_4$) to name a few [MESCHKE, KUGEL, HIDAKA, GOPALAN, ROSENMAN]. The study of BT-based ceramics with stoichiometric compositions different from pure BT has become one of the most important subjects of ferroelectrics in recent years. Particularly, substitution of Sr^{2+} ions in place of Ba^{2+} ions into BT leads to a solid solution, barium strontium titanate (BSTx or Ba$_{(1-x)}$Sr$_x$TiO$_3$, where $0 \leq X \leq 1$). Ferroelectric materials have been synthesized by various techniques, the most commonly used today being the technical or sol-gel process for the production of powders or thin films [BOLAND, PARK, ZHU, KAMALASANAN]. Another technique used to obtain powders is hydrothermal process [XU, RAZAK, VOLD, CHENG]. Finally there is the conventional route solid-state reaction of mixed oxides [VITTAYAKORN, IANCULESCU, CHAISAN] to obtain powders and solid ceramics. The interest of processing highly dense BSTx ceramics is that the Curie temperature and thus the dielectric properties can be tuned using the chemical variations between SrTiO$_3$ and BaTiO$_3$ [BERBECARU, YUN].

This chapter provides the description of an alternative solid-state reaction route based on high energy ball milling and subsequent sintering for the synthesis and densification of BSTx bulk ceramics. It provides a more direct route than the conventional route of mixed oxides. In addition to presenting structural characterization and results of electrical measurements (dielectric constant versus temperature curves and ferroelectric hysteresis loops), a novel technique known as contact resonance piezoresponse force microscopy (CR-PFM) is applied in the detection and characterization of ferroelectric domains in the BSTx samples.

2. Experimental procedures

Titanium oxide (TiO$_2$ - 99.9% purity, anatase phase, Aldrich), barium carbonate (BaCO$_3$ - 99.9% purity, Aldrich) and strontium carbonate (SrCO$_3$ - 99.9% purity, Aldrich) powders were weighed according to the stoichiometric proportion of Equation 1. The mixture was ball-milled in a high energy vibratory mill (SPEX 8000) for 6 h using a nylamid vial with 10 toughened zirconia balls (10 mm diameter) and a ball-to-powder weight-ratio of 10:1. The milled powders were uniaxially pressed in a stainless steel cylindrical die (10 mm inner diameter) using 1 GPa pressure. The green compacts were placed on top of TiO$_2$ substrates inside alumina crucibles (to avoid the reaction of barium oxide with the crucibles) and reactively sintered using a two-step heating program (Figure 1). The compacts are heated from room temperature up to 1273 K (reaction temperature) at 5 K/min and held there for 1 h. The temperature is then increased to between 1523 and 1573 K (sintering temperature) at 3.0 K/min and held there for 2 h. Finally they are cooled to room temperature at 3.0 K/min. The resulting BSTx sintered ceramics with the perovskite ABO$_3$ structure complied with the formula Ba$_{(1-x)}$Sr$_x$TiO$_3$ where $X = 0, 0.1, 0.2 \dots 1$ and were named accordingly, e.g. BST3 refers to the $X = 0.3$ composition. The starting powders, milled powders and sintered ceramics were characterized by x-ray diffraction (XRD) using a Rigaku Dmax-2100 diffractometer equipped with Co Kα radiation; scanning electron microscopy (SEM) using a

Philips XL30 ESEM; transmission electron microscopy (TEM) using a JEOL 2010; micro-Raman scattering using a DILOR unit; differential scanning calorimetry (DSC) using a Mettler Toledo; and thermogravimetric analysis (TGA) using a SDTA851 Mettler Toledo. The bulk density of the sintered ceramics was determined using the Archimedes method.

$$(1-X)BaCO_3 + XSrCO_3 + TiO_2 \rightarrow Ba_{(1-x)}Sr_xTiO_3 + CO_2 \text{ (g)} \tag{1}$$

This fabrication method is a modification of the conventional route (solid state reaction) for the manufacture of $Ba_{(1-x)}Sr_xTiO_3$ ceramics. It requires less processing steps (Figure 2), is more straightforward than other chemical processes and, most important, leads to the fabrication of high-density ceramics with very low porosity.

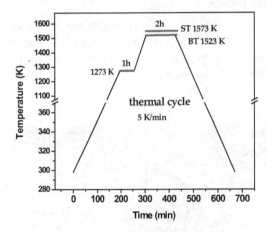

Fig. 1. Heat treatment used for the manufacture of BSTx.

Alternative Route

Milling

Uniaxial pressing

Heat treatment
(reaction-sintering)

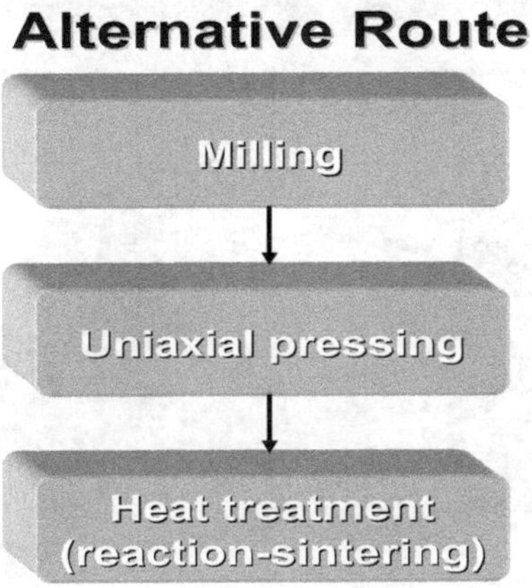

Fig. 2. Stages of the proposed alternative route for the manufacture of BSTx.

It must be noted that the high energy milling process used allows for homogenization and particle size reduction of the starting powders. It is a more efficient process than conventional attrition, planetarium or automatic-agate milling. The milled powders were compacted using much higher pressure (~1.0 GPa) than applied using the conventional approach (~0.1 GPa). This resulted in uniform green compacts which can conduct an homogeneous chemical reaction. Finally, the thermal treatment induces simultaneous reaction and sintering (reaction-sintering, Figure 1) as opposed to the conventional manufacturing process where the starting powders are milled, thermally-treated to react, a second milling process is performed, and then the twice-milled powders are pressed into compacts. Finally, a second thermal treatment (sintering) densifies the compacts.

3. Results and discussion

3.1 Fabrication of BSTx powders

The high energy milling process significantly affects the size and shape of the starting powders, as seen in the SEM micrographs of Figure 3. Figure 4 shows the TEM micrographs of the starting powders after milling for 6 h. A final particle size of less than 50 nm was attained for both extreme concentrations (only $BaCO_3$ and TiO_2, and only $SrCO_3$ and TiO_2).

Fig. 3. SEM image of starting powders (a) milled for 6 hours, and (b) before milling.

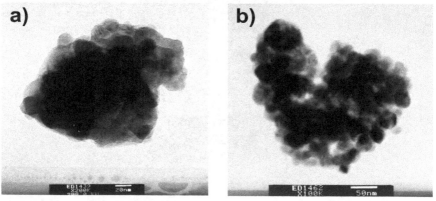

Fig. 4. TEM micrographs of $(1-X)BaCO_3 + XSrCO_3 + TiO_2$ powder milled for 6 hours, with (a) X = 0.0 and (b) X = 1.0.

The x-ray diffraction patterns for the milled $BaCO_3$, TiO_2 and $SrCO_3$ powder mixtures are presented in Figure 5. The main $BaCO_3$ peak, located at approximately $2\theta = 28°$ is present up to X = 0.9, and the small peak at approximately 30° belongs to TiO_2. As expected, increasing the $SrCO_3$ content increases the height of the $SrCO_3$ peak located at approximately 30°. There is a degree of amorphization due to the creation of defects in the crystal structure. Figure 6 presents the thermogravimetric analyses (TGA) curves for the same collection of samples as analyzed by x-ray diffraction (Figure 5). All the curves are similar in their key characteristics. The weight loss behavior for a single sample, with X = 0.65, is shown in Figure 8. There are four stages of weight loss centered at 403, 773, 973 and 1273 K. In the first stage, from room temperature (RT) up to 403 K, a weight reduction of approximately 1.3% occurs due to evaporation of water from the material surface. In the second stage, from 403 to 773 K, the weight loss of about 5.2% is related to the loss of chemically bound water in the form of OH groups from the BaOH that was formed when BaO combined with water during milling [BALÁZ]. This loss typically occurs between 473 and 873 K [ASIAIE]. The phenomenon of water loss has also been reported to occur in other carbonates between 593 and 723 K [DING]. The third stage, between 773 and 973 K, is not related to any structural modification of the BSTx samples. According to the literature, the decomposition of strontium and barium carbonates to form CO_2 occurs at higher temperatures, between 1023 and 1273 K [JUDD, L'VOV, MAITRA]. However, in this case, the generation of CO_2 begins as early as at 833 K and runs up to 1273 K. The x-ray difffraction patterns of Figure 7 show the appearance of a peak ($2\theta \approx 36°$) at 873 K, which corresponds to the formation of the perovskite structure of $BaTiO_3/SrTiO_3$. Therefore, the weight loss from 773 to 1323 K can be considered a single stage that varies with Sr content. It is directly related to the CO_2 excess from the carbonates used as starting powders (see Equation 1). It is the difference between the weight of the $(1-X)BaCO_3 + XSrCO_3 + TiO_2$ starting powders and that of the resulting $Ba_{(1-X)}Sr_XTiO_3$. Table 1 lists the total weight loss, the loss in the different stages, and the weight loss expected from CO_2 liberation. For the group of samples as a whole, an approximate 5% weight loss was observed between RT and 773 K, and the weight loss from 773 to 1323 K corresponds closely to the stoichiometric CO_2 loss.

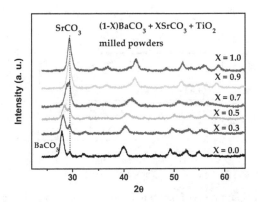

Fig. 5. X-ray diffraction patterns for $(1-X)BaCO_3 + XSrCO_3 + TiO_2$ milled powders.

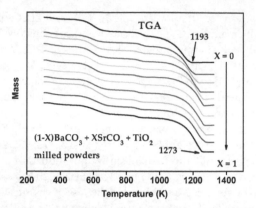

Fig. 6. Thermogravimetric curves of (1-X)BaCO$_3$ + XSrCO$_3$ + TiO$_2$ milled powders.

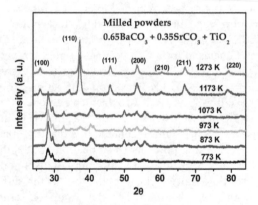

Fig. 7. XRD patterns of 0.65BaCO$_3$, 0.35SrCO$_3$ and TiO$_2$ powders milled for 6 hours and subsequently heat treated at different temperatures.

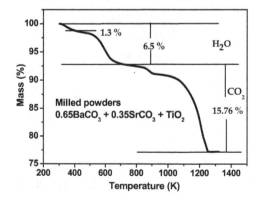

Fig. 8. Thermogravimetric analysis of 0.65BaCO$_3$, 0.35SrCO$_3$ and TiO$_2$ powders milled for 6 hours.

The thermogravimetric plots also show a correlation between the Sr content and the temperature at final weight loss (zero slope section). For example, the BST0 sample stops losing weight at around 1193 K but the BST10 (with highest Sr content) at around 1273 K. The temperature at which no more weight loss is observed marks the end of the reaction, as corroborated by the XRD measurements (Figure 7). The thermogravimetric curves can therefore be used as guides to establish the reaction temperature in the BSTx system.

Sample ID	% Total weight loss	% Weight loss from RT to 773 K	% Weight loss from 773 to 1323 K	% Weight loss CO$_2$ stoichiometric
BST0	20.43	5.56	14.87	15.87
BST1	19.38	4.12	15.26	16.16
BST2	21.02	5.48	15.54	16.47
BST3	21.59	5.61	15.98	16.78
BST4	21.52	5.94	15.58	17.10
BST5	21.39	5.80	15.59	17.44
BST6	21.76	4.85	16.91	17.79
BST7	22.48	5.19	17.29	18.15
BST8	24.29	5.76	18.53	18.53
BST9	23.46	5.42	18.04	18.93
BST10	22.87	4.52	18.35	19.34

Table 1. Thermogravimetric analysis of (1-X)BaCO$_3$ + XSrCO$_3$ + TiO$_2$ powder after high energy milling for 6 hours.

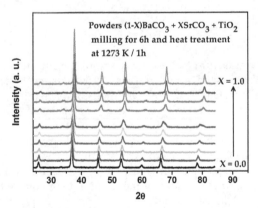

Fig. 9. X-ray diffraction patterns of (1-X)BaCO$_3$ + XSrCO$_3$ + TiO$_2$ powder milled and heat treated at 1273 K for 1 h. The crystal structure is perovskite-type ABO$_3$.

The diffraction patterns for the range of samples after milling and thermal treatment at 1273 K for 1 hour are shown in Figure 9. All the samples present the perovskite-type structure ABO$_3$, indicating the reaction was successfully completed. The formation temperature of the BaTiO$_3$ phase coincides with that reported by L. B. Kong [KONG], where the rutile phase of titanium oxide (TiO$_2$) was used instead of TiO$_2$ anatase phase. The peaks shift to higher angle as the Sr content increases. For example, the most intense BaTiO$_3$ peak in Figure 10 shifts from 36.79° for the BaTiO₃ (BST0) to 37.77° for SrTiO₃ (BST10). This shift corresponds to a reduction in unit cell size, consistent with the difference between the ionic radii of Ba^{2+} (1.34 Å) and Sr^{2+} (1.12 Å). The ABO_3 perovskite structure refers to the relative position of the Ba^{2+}, Sr^{2+} and Ti^{4+} ions with respect to oxygen (O^{2-}). The structure can present as different phases, depending on the material and temperature. For the BSTx system it may be cubic or tetragonal at room temperature, depending on the strontium content. In this case, we observed the cubic phase after 1 hour of heat treatment at 1273 K (Figure 9). The relationship between the cubic and tetragonal phase and the volume of the unit cell will be discussed in more detail when disucssing compaction.

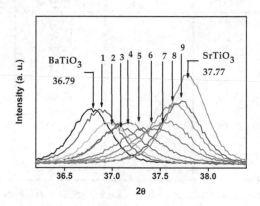

Fig. 10. Expanded view of the primary peak in Figure 9, showing the shift in peak position with increasing Sr content from X = 0 (BaTiO₃) to X = 1 (SrTiO₃).

3.2 Sintering

During conventional manufacture (solid-state reaction) the starting oxide powders are first milled and calcined. The reacted powders (having the desired stoichiometry) are milled once again to reduce the particle size. They are then compacted into disks or other shapes (green ceramic) and finally, thermally treated to sinter the compact. At the sintering temperature, the ceramic particles coalesce with each other to form grains, the material shrinks and the pores are eliminated [KINGERY]. The alternative manufacturing route of the present work is more straightforward. The starting powders are milled only once to homogeneize and reduce the particle size before being compacted into disks or other shapes. The green ceramics are thermally treated at the reaction temperature (1273 K) to obtain the perovskite-type structure ABO_3. A second heating step is subsequently applied to sinter the ceramic (Figure 1). In this way, dense ceramics (>90% of the theoretical density) are obtained. The sintering temperatures of the samples with intermediate stoichiometry varied between 1523 K for pure BaTiO₃ and 1573 K for pure SrTiO₃. The temperatures were empirically determined. Such reduced sintering temperatures can only be applied to powders with particle size smaller than 50 nm, which is uniquely attained in our alternative route, thanks to the high energy milling. Whenever the BaTiO₃ samples were sintered at temperatures higher than 1523 K, severe strain was observed as the melting temperature of the material (1898 K [PRADEEP]) was approached. Ceramic shrinkage, or pore reduction, is directly related to the initial particle size, as shown in Equation 2 [KINGERY]:

$$d - d_0 = (2k)^{1/2} t^{1/2} \tag{2}$$

where d and d_0 are the initial and final grain diameters, respectively, k is a constant, and t is the processing time. Sintering temperatures for the conventional route of ceramic preparation are reported to be between 1623 and 1703 K [ZHONG, WODECKADUS, TERANISHI]. The x-ray diffraction patterns of the thermally treated BSTx ceramics (Figure 11) clearly show the perovskite-type structure. As in the case of the powders, the peaks shift to higher angle as the Sr content increases. From Figure 12 it can be seen that the strongest peak shifts linearly (~0.1°) for each 0.1 increase in X (Sr content), and thus can be used to determine the sample composition (stoichiometry). Figure 13 plots the peak position as a function of the Sr content. All the thermally treated BSTx ceramics have the correct stoichiometry and present the perovskite-type structure ABO$_3$. However, the structure can be either tetragonal or cubic at room temperature depending on the processing conditions. For the BaTiO$_3$ the relative position of the (002) and (200) peaks determine the tetragonality of the phase, i.e., the ratio of the lattice parameters c/a. Such peaks appear at approximately $2\theta = 53°$.

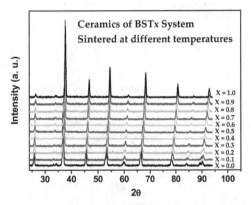

Fig. 11. X-ray diffraction patterns of BSTx ceramics system sintered from 1523 to 1573 K for 2h.

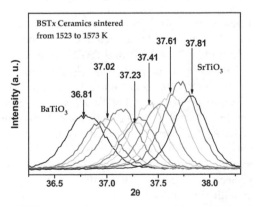

Fig. 12. Main XRD peaks of BSTx ceramic system sintered from 1523 to 1573 K of the figure 11.

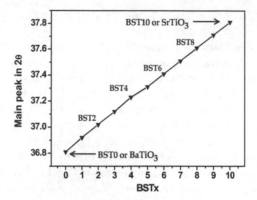

Fig. 13. Main diffraction peak position (Figure 12) as a function of stoichiometric composition.

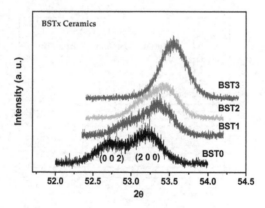

Fig. 14. Separation (or lack therof) of the (002) and (200) reflections of sintered ceramics BST0, BST1, BST2 and BST3 providing a measure of the c/a ration.

The phase is cubic and paralectric if the c/a ratio = 1. In such a case there is only a single peak near $2\theta = 53°$. If c/a > 1 the phase is tetragonal and ferroelectric. In such case there is a double peak. Figure 14 shows the BST0, BST1 and BST2 samples have a double peak, while sample BST3 apparently has only a single peak. Rietveld refinement (using Maud Program) was performed on the diffraction patterns of Figure 11 in order to determine the unit cell lattice parameters. The results, presented in Figure 15, show a gradual decrease in lattice parameters with increasing Sr content. This is due to the substitution of strontium Sr^{2+} ions (with ionic radius of 1.12 Å) for barium Ba^{2+} ions (with ionic radius of 1.34 Å). The tetragonality (c/a) decreases also from 1.008 (BST0) to 1.0014 (BST3) while for BST4 to BST10 it has value of 1. Differential scanning calorimetry (DSC), Raman spectroscopy, as well as the ferroelectric and dielectric measurements confirmed this phase transition, and are presented below, along with the cubic-to-tetragonal phase transition temperature (Curie temperature).

Ceramic Sample	Position in 2θ	Maximun intensity of the peak	Peak width	Lattice parameter of unit cell	Tetragonality c/a
BST0	52.694	704	0.485	a = 3.9953	1.0089
	53.227	1265	0.543	c = 4.0311	
BST1	52.942	388	0.520	a = 3.9844	1.0067
	53.368	1190	0.487	c = 4.0114	
BST2	53.124	490	0.446	a = 3.9785	1.0048
	53.459	1185	0.477	c = 3.9979	
BST3	53.457	1327	0.335	a = 3.9684	1.0014
	53.611	1764	0.360	c = 3.9737	
BST4	53.725	10	0.146	a = 3.9611	1

Table 2. Unit cell parameters and fitting parameters of peaks around 53° in 2θ.

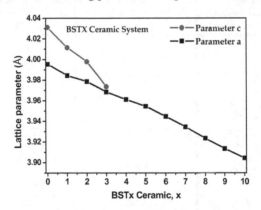

Fig. 15. Lattice parameter as a function of Sr content.

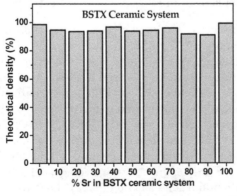

Fig. 16. Percentage of theoretical density of sintered compacts as a function of Sr content.

3.3 Density

With the actual volume of the unit cell and considering the number of barium, strontium, titanium and oxygen atoms composing the ABO_3 unit cell, the theoretical density can be calculated using:

$$\text{Theoretical Density} = \frac{\text{Unit cell mass}}{\text{Unit cell volume}} \tag{3}$$

The unit cell volume ($a^2 \times c$) was calculated using the results of the Rietveld analysis. The mass of the unit cell is calculated considering 3 oxygen, 1 titanium, (x) strontium and (1-x) barium atoms. The bulk density of the samples was determined by the Archimedes method. Figure 16 plots the bulk density as percentage of the theoretical density. All samples have bulk densities higher than 90% of theoretical, proving that the alternative fabrication route can attain a high densification of the ceramics. The density caluclation and measurment results are presented in Table 3.

BSTx Sample	Unit cell volume ($Å^3$)	Theoretical density ($g\ m^{-3}$)	Bulk density ($g\ m^{-3}$)	% Theoretical density
BST0	64.349	6.016	5.928	98.526
BST1	63.684	5.950	5.630	94.632
BST2	63.283	5.857	5.486	93.661
BST3	62.675	5.782	5.430	93.910
BST4	62.152	5.698	5.513	96.758
BST5	61.839	5.593	5.246	93.798
BST6	61.377	5.501	5.191	94.364
BST7	60.905	5.408	5.191	95.986
BST8	60.393	5.317	4.883	91.838
BST9	59.934	5.220	4.755	91.080
BST10	59.514	5.118	5.078	99.217

Table 3. Theoretical and bulk (measured) Densities of BSTx ceramic system.

SEM micrographs of transversely fractured sections from the BST0, BST4, BST8 and BST10 samples are shown in Figure 17. The BST0 ($BaTiO_3$) and BST10 ($SrTiO_3$) ceramics have a uniform compact morphology, present no cracks and have low porosity. The grain boundaries are not observable. Samples BST0 and BST10 have a bulk density of 98.52% and 99.21% of the theoretical, respectively. Even with the lower-than-conventional sintering temperatures, it is possible that the observed morphology resulted from liquid phase formation due to the small particle size [KINGERY, BARSOUM, BARRY]. Grain size distributions for BST0 and BST10 are shown in Figure 18. For the BST0 sample, grain sizes between 1 and 3.5 μm (2 μm average) were measured, while for the BST10 sample, sizes ranged ranged from 1 μm to 2.6 μm (1.4 μm average). The BST4 and BST8 samples, consisting of a barium-strontium solid solution (Ba,Sr)TiO_3, have a distinctly smaller grain size, with distributions of 0.2 to 1.3 μm (650 nm average) and 0.3 to 0.9 μm (550 nm average), respectively. Generally speaking, the conventional route for fabrication of mixed oxide ceramics leads to grains larger than 1 μm, in most cases on the order of several microns [LIN, LIOU, LU]. Smaller grains are only attained with nanopowders and special sintering processes such as spark plasma sintering (SPS) [DENG] or hot pressing [XIAO]. Other methods for inhibiting grain growth include the incorporation of 1 wt% of Na, Mn or Mg ions [LIOU].

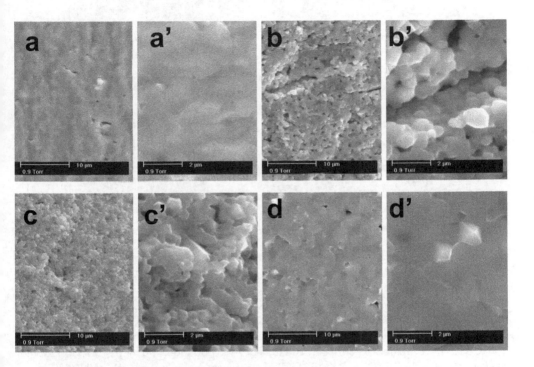

Fig. 17. SEM micrographs of transversely fractured BST0, BST4, BST8 and BST10 samples at two magnifications: (a) BST0 sintered at 1523 K, (b) BST4 sintered at 1543 K, (c) BST8 sintered at 1573 K, (d) BST10 sintered at 1573 K.

3.4 Curie temperature (Tc) via differential scanning calorimetry

DSC measurements were conducted from 203 to 423 K in a nitrogen atmosphere with a heating rate of 20°C/min. Figure 19 presents the DSC curves for samples BST0 to BST3. The cubic to tetragonal phase transition (Curie temperature, Tc) is an endothermic event. Tc decreases with increasing content of Sr in the samples, i.e., as the Sr^{2+} ions replace Ba^{2+} ions. This behavior was previously reported by Rupprecht and Bell [RUPPRECHET].

The linear dependence between the Tc and at.% Sr is described by Equation 4, the result of fitting the experimental data (Figure 20):

$$Tc = 128.4871 - 31.469 * X \qquad (4)$$

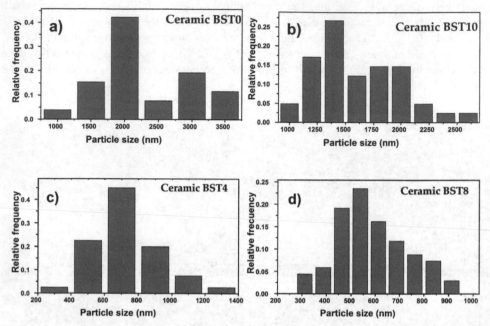

Fig. 18. Grain size distribution of (a) BST0, (b) BST10, (c) BST4 and (d) BST8 ceramic samples.

where Tc is the Curie temperature and x is the Sr content in at.%. Table 4 presents the Curie temperatures of samples BST0 to BST3 determined by Equation 4. The equation was extrapolated to the composition of samples BST4 to BST10. The determined Curie temperatures resulted not so different from those reported in the literature, for example, the BST35 system (35 mol% of Sr) has an approximate Tc of 292 K [ALI] compared to 291.35 K determined with Equation 4. The BST3 system was reported to have a Tc of 306~307 K [PITICESCU], compared to 307.08 K calculated with Equation 4. The $Ba_{60}Sr_{40}TiO_3$ system (BST4) has a Tc of 272 K (274.15 K) [FETEIRA], compared to our calculation of 276.6 K.

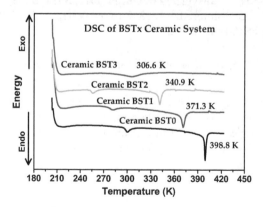

Fig. 19. Differential scanning calorimetry curves of the BST0, BST1, BST2 and BST3 samples.

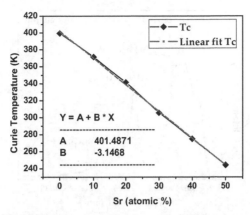

Fig. 20. Linear fit of the Curie temperature (Tc) from the DSC curves.

Sample ID	Calculated Tc (K)	Experimental Tc (K)
BST0	401.48	399.80
BST1	370.02	371.36
BST2	338.55	340.96
BST3	307.08	306.62
BST4	275.61	273.0
BST5	244.15	241.65
BST6*	212.68	
BST7*	181.21	
BST8*	149.74	
BST9*	118.27	
BST10*	86.80	

Table 4. Experimental Curie temperatures (via DSC), and those calculated using Equation 4. Tc for samples BST6 – BST10 were not experimentally determined due to exceeding the temperature range of the differential scanning calorimeter.

3.5 Curie temperature (Tc) via Raman spectroscopy

Figure 21 shows the Raman scattering spectrum (radiation wavelength = 514.5 nm) for BaTiO₃ (BST0) with perovskite-type structure ABO_3 and tetragonal phase at room temperature. Raman active phonons for the P4/mmm tetragonal symmetry are represented by 3A1 + B1 + 4E. Long-range electrostatic forces induce a splitting in the transverse and longitudinal phonons, resulting in a split of the Raman active phonons represented by 3 [A1 (TO) + A1 (LO)] + B1 + 4 [E (TO) + E (LO)] [SHIRATORI]. Raman shift bands are reported at 250, 520 and 720 cm⁻¹ with a sharp peak at around 306 cm⁻¹ [DIDOMENICO, ROUSSEAU, BASKARA].

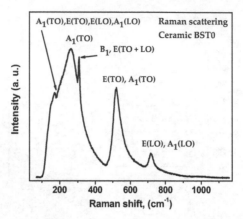

Fig. 21. Raman scattering spectra for BST0.

The shoulder at around 180 cm^{-1} in bulk BaTiO$_3$ is attributed to the coupling of the three disharmonic phonons A1 (TO) [VENKATESWARAN, FREY]. Figure 23 shows the Raman scattering spectra of the BST0 sample at different temperatures. The 250, 520 and 720 cm^{-1} bands as well as the 306 cm^{-1} peak decrease gradually as the temperature increases. At 403 K, the sharp 306 cm^{-1} peak disappears, indicating the transition from cubic to tetragonal phase (Tc). The transition temperature was previously reported by C. H. Perry [PERRY]. Thus, the sharp peak around 306 cm^{-1} indicates whether the BSTx system is in the tetragonal or cubic phase. The relative intensity of the 306 cm^{-1} peak as a function of temperature for BST0 is presented in Figure 22. Figures 24 and 25 show the temperature-dependent Raman scattering spectra for BST1 and BST2, respectively. The cubic to tetragonal phase transition is observed in the range from 363 to 373 K for BST1 and 333 to 343 K for BST2. Both ranges match those determined by DSC. Figure 26 shows the Raman scattering spectra for BST1 to BST5 at room temperature. Samples BST4 and BST5 do not present the sharp peak at 306 cm^{-1}, i.e., they have a stable cubic phase and a tetragonality of 1 (c/a = 1). These results are consistent the XRD and DSC results presented earlier.

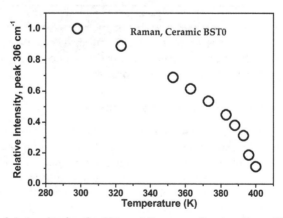

Fig. 22. Realtive peak intensity for the 306 cm^{-1} Raman reflection (from Figure 21) as a function of temperature for BST0.

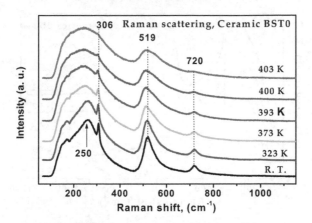

Fig. 23. Raman scattering spectra at different temperatures for BST0.

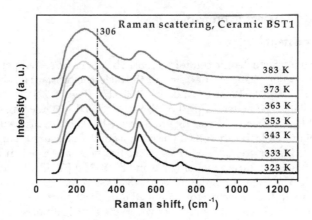

Fig. 24. Raman scattering spectra at different temperatures for BST1.

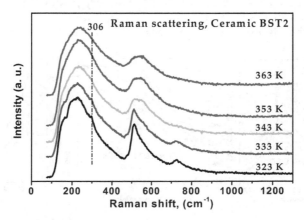

Fig. 25. Raman scattering spectra at different temperatures for BST2.

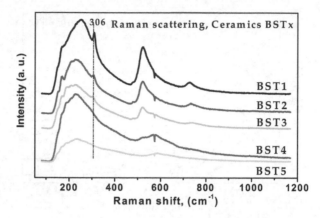

Fig. 26. Raman scattering spectra at room temperature for BST1, BST2, BST3, BST4 and BST5.

3.6 Dielectric and ferroelectric properties

3.6.1 Dielectric constant

When a polycrystalline ferroelectric ceramic is cooled below its Curie temperature, some of its properties undergo strong changes. For example, the dielectric constant (ε) shows a maximum at Tc for BST0, BST1 and BST3 at 0.1, 1.0, and 100 kHz (Figure 27), a typical ferroelectric behavior of perovskite-type materials [MILLAR, STANFORD]. A widening of the peak at Tc has been reported to occur with decreasing grain size [KINOSHITA, SAKABE]. The dielectric constant has a magnitude larger than 1000 within most of the measured temperature interval, and it decreases at higher frequencies. This effect has been observed in BaTiO$_3$ ceramics containing Zr [DEB] and in those containing Na and Bi (BTNx) [GAO]. The dielectric constant has polar and ionic parts, therefore, the dielectric dispersion can be attributed to the dipoles ceasing to contribute to the dielectric constant as the frequency increases [MERZ, 1954]. The dielectric relaxation effect occurs at frequencies where the electric dipoles can no longer follow the oscillation of the applied electric field. The relaxation frequency can be determined by a drop in the real part of ε and a maximum in the imaginary part [RAVEZ]. Although there is dielectric relaxation in the BSTx samples, the material does not present typical reflexor behavior. That is, there is no change in the position of Tc when the frequency changes. Another

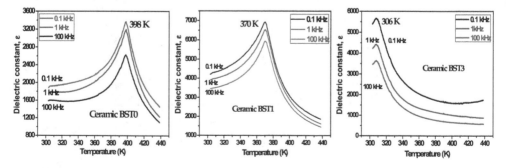

Fig. 27. Dielectric constant curves for (a) BST0, (b) BST1 and (c) BST3.

ferroelectric material with such a behavior is Pb(Mg$_{1/3}$Nb$_{2/3}$)O$_3$ [KOVALA]. The determined Tc's agree well with those obtained by DSC and Raman. The dielectric constant peak values are 3.179, 6.540 and 4.432 for BST0, BST1 and BST3 ceramics respectively, 2 to 3 orders of magnitude larger than those of other materials conventionally used in capacitors or CMOS (complementary metal oxide semiconductor) devices [ROBERTO, WILK].

All of the BSTx samples showed a maximum in the dielectric constant at Tc. Above this temperature, the dielectric behavior obeys the Curie-Weiss law and has the form [BURFOOT]:

$$\varepsilon = \frac{C}{T - T_0} \qquad (5)$$

where ε is the dielectric constant (material permittivity), C is the Curie-Weiss constant, T is the temperature of the material and T_0 is the Curie-Weiss temperature.

3.6.2 Ferroelectric hysteresis loops

When an external electric field E is applied to a dielectric material, it produces a P vs. E curve. In the case of ferroelectric materials there is a delay in the P response to the E stimulus, i.e., hysteresis. A freshly manufactured ferroelectric has a zero spontaneous net polarization ($P_s=0$). When an external electric field is applied, nucleation and growth of the ferroelectric domains occur [MERZ, 1954].

The shape of the ferroelectric curve $P = f(E)$ depends on both time and temperature. In the present work, ferroelectric measurements were performed at room temperature (~298 K) at a fixed frequency of 100 Hz using a comercial Sawyer-Tower circuit (ferroelectric RADIANT Test System) [SAWYER and TOWER]. Figure 28 presents the hysteresis loops for BST0, BST1, BST2 and BST3. The **P-E** urves exhibit the typical behavior of polycrystalline ferroelectric ceramics [MERZ, 1953]. The remanent polarization (P$_r$) is low compared to that of BaTiO$_3$ single crystals (>20 μC/cm^2 [SRIVASTAVA]), but higher than that of nanocrystalline BaTiO$_3$ (P$_r$ < 1 μC/cm^2 [BUSCAGLIA]). It is comparable to that of BaTiO$_3$ ceramics with grain sizes of around 1 μm [TAKEUCHI].

3.7 Piezoresponse Force Microscopy (PFM)

3.7.1 Ferroelectric domain observation

Ferroelectric materials are composed of ferroelectric domains, which can be observed by polarized light microscopy [RUPPRECHT, ARLT], scanning electron microscopy [CHOU, ROUSSEAU] and transmission electron microscopy [FREY, GANPULE]. To be detected by these techniques, some sort of chemical attack is necessary to reveal the ferroelectric domains, as they demonstrate a preferential rate of erosion [FETEIRA, LAURELL]. Furthermore, these techniques do not directly indicate the direction of polarization (direction of the domain); they only discriminate one domain from another. Piezoresponse force microscopy (PFM) does not suffer from these shortcomings [RABE], allowing visualization of ferroelectric domains with sizes on the order of 1 μm [SAURENBACH, WITTBORN], accurate detection of the polarization direction [ENG 1998, CHO] and reconstruction of the three-dimensional orientation of the domains [ENG, 1999]. Figure 29 is a diagram of oriented domains in ferroelectric grains. Figure 29 (a shows a domain up and one down (antiparallel ↑↓), separated by domain walls (gray stripe). The adjacent domains are oriented in opposite directions (180°). If these domains are observed from above, we would see that the direction of the domains are perpendicular to the surface, either pointing down into the sample (crosses) or up out of the sample (circles)., Both types are described as out-of-plane (OOP) (Figure 30). Figure 29 (b shows domains oriented at 90° (↑→) relative to one another. If these domains are observed from above, we would see domains in the plane of or parallel to the surface. These dominains are described as in-plane (IP) (Figure 30).

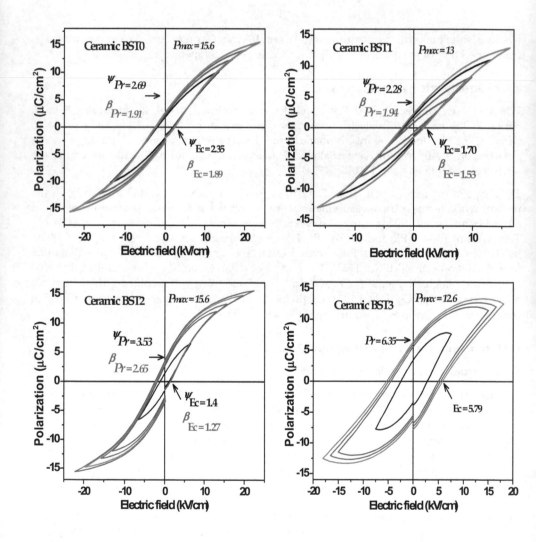

Fig. 28. P-E loops obtained for BST0, BST1, BST2 and BST3.

It isimportant to note that Figure 29 represents the ideal case of monocrystal domains ideally oriented, i.e., the direction of the observed polarization vectors are orthogonal. For the real case of a polycrystalline ceramic, the direction of the polarization vector would be random, i.e., a domain can point in any direction in 3D space. Therefore, the signals from piezoresponse measurements in OOP or IP are the projections of these random polarization vectors.

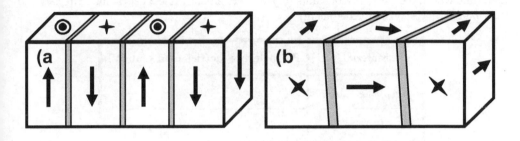

Fig. 29. Ferroelectric domains with differents orientations of the polarization vector. (a out-of-plane (OOP) and (b in-plane (IP).

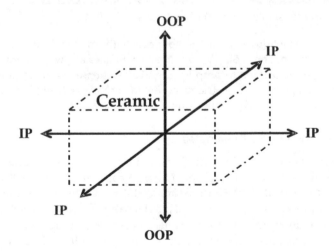

Fig. 30. Standard terminology for polarization vector orientations relative to the observed surface in piezoresponse force microscopy (PFM). IP refers to in-plane, and OOP is out-of-plane.

3.7.2 Contact Resonance-Enhanced Piezoresponse Force Microscopy (CR-PFM)

In addition to the conventional PFM technique there is another mode called contact resonance-enhanced PFM or CR-PFM [HARNAGEA]. R-PFM is based on the same principles of operation of as PFM, the only difference being the value of the frequency of the alternating voltage applied to the tip (V_{AC}). For contact resonance frequencies (CR-PFM), the strain amplitude response of the material can be more than one or two orders of magnitude higher than the amplitudes recorded by conventional PFM (quality factor Q) [HARNAGEA]. In other words, R-PFM is more sensitive than PFM and can be applied to materials whose piezoelectric constants (d_{ij}) are very small. Table 5 lists some materials and their piezoelectric constant (coefficient) .

Material	Piezoelectric coefficient, d_{33} (pm/V)
$Sr_{0.61}Ba_{0.39}Nb_2O_6$	200
$BaTiO_3$	190
PZT4	291
PZT5a	373
Quartz	3

Table 5. Piezoelectric constants of different materials.

The piezoelectric constant (d_{33}) indicates how much the material will be deformed (in picometers) for each applied volt (V) applied. or Fmaterials with a low piezoelectric constant, it would be necessary to amplify the piezoresponse signal by applying a high voltage, risking a change in piezoelectric response. In such cases, R-PFM is an alternative to observe ferroelectric domains while still using a low voltage.

3.7.3 Results of R-PFM for BSTx samples

R-PFM measurements for the BSTx samples were made using the modified conventional method on an atomic force microscope (AFM, Veeco di DimensionTM 3100), usig conductive tips of Cr/Pt (Budget Sensors Tap I300E) with a force constant k = 40 N/m, and a free-resonance frequency was 300 kHz. The samples were polished starting with No. 500 sandpaper down to 0.3 μm alumina. The samples were not attacked by any chemical or mechanochemical process, insuring that topography did not contribute to the piezoresponse signal. AFM images of the contact resonance mode piezoresponse (CR-PFM) for BST0, with a grain size of 1-2 μm, are shown Figure 31. The three images, taken of the same area and at the same time, show (a) the topography, (b) the OOP piezoresponse amplitude, and (c) the OOP piezoresponse phase. Measurements were performed over a 5 x 5 μm area, using an applied voltage was 2V and a contact resonance frequency of 1350 kHz. The amplitude and phase piezoresponse images present different characteristics. In the Figure 31b, the piezoresponse amplitude image allows us to visualize domains that leave or enter of the surface (OOP), where the amplitude A can be the same for antiparallel dominions ($\uparrow\downarrow$). Different regions are distinguished by their gray tones, delimited by black contours, which correspond to the ferroelectric domain walls. The domain walls are not observed in the topography, as that would have required etching (chemical attack) [FETEIRA]. Regions that appear as similar shades of gray in the piezoresponse amplitude image (Figure 31b) appear as contrasting bright and dark in the piezoresponse phase image (Figure 31c). This indicates a phase shift of 180° from one domain to another, but with the same piezoresponse amplitude.

These results show that the R-PFM images of amplitude and phase are not influenced by the topography of the samples. In contrast, standard measurements of piezoresponse force microscopy (PFM) were performed at 20 kHz with an alternating voltage (V_{ac}) of 2-15 V without obtaining a piezoelectric response. A piezoresponse was obtained with contact resonance frequencies (~1350 kHz), with a quality factor Q betwen 50 and 100. Taking into account both the quality factor Q and the piezoresponse signal reported about 190 pm/V and to 419 pm/V [SHAO] for the case of BaTiO$_3$, we can estimate a small piezoelectric constant of our material is of the order of units or tens of pm/V.

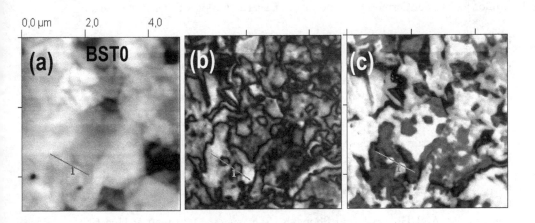

Fig. 31. R-PFM of BST0 at a frequency of 1.350 MHz and an excitation voltage of 2 V. (a) topography, (b) piezorepsonse amplitude, and (c) piezoresponse phase.

4. Conclusion

The mixed oxides modified route (solid-state reaction) is a direct alternative to obtain highly densified BSTx ceramics. In this route, the high-energy ball milling and the applied heat treatment allowed the preparation of nanometric powders (less than 200nm) with Perovskite-type structure ABO_3. The Curie temperature of the BSTx ceramics was unambiguosly determined as a function of temperature by several techniques: Raman spectroscopy, differential scaning calorimetry, and measurements of dielectric constant. This temperature was succesfully tuned from 87 K to 400 K by varying the Sr/Ba ratio, as expected. However, a shifting in the orthorhombic to tetragonal phase transition was observed in the sintered ceramics. For instance, the OTPT for the BST0-BST2 samples was shifted down 17-27 K with respect to the literature. The origin of this shifting is probably residual stresses associated to the fine-grained microstructure of the sintered samples.

The group of BST0, BST1, BST2 and BST3 ceramics present P(E) curves with ferroelectric behavior at room temperature. The other cases present paraelectric behavior. Moreover, the BST0, BST1and BST2 samples present rather low maximum and remanent polarization and coercive fields. BSTx ceramics (x = 0, 1, 2, 3) displayed piezoelectric response in the contact resonance piezoresponse force microscopy mode (CR-PFM). The polycrystalline BSTx ceramics showed ferroelectric domains with sizes several times smaller than the grains.

5. Acknowledgment

This work was partially supported by CONACYT, Mexico. The authors would like to thank Michael Boldrick Ph. D. and Rodrigo Mayen Mondragón Ph. D. for their help regarding the english translation. Besides, we thank Pedro García J., J. Eleazar Urbina A., M. Adelaido Hernández L., Francisco Rodríguez M., Agustín Galindo S., Rivelino Flores F., Alfredo Muñoz S., Ma. del Carmen Delgado C. and Eduardo Larios for their technical aid.

6. References

Ali, N. J. & S. J. Milne, "Comparison of powder synthesis routes for fabricating $(Ba_{0.65}Sr_{0.35})TiO_3$ ceramics", *J. Mater. Res.*, 21 (2006) 1390-1398.

Arlt, G. & P. Saxo, "Domain configuration and equilibrium size of domain in $BaTiO_3$ ceramics", *J. Appl. Phys.*, 5 (1980) 4956-4960.

Asiaie, R.; Weidong Zhu, Sheikh A. Akbar, & Prabir K. Dutta, "Characterization of submicron particles of tetragonal $BaTiO_3$", *Chem. Mater.*, 8 (1996) 226-234.

Baláz, P. & B. Plesingerova, "Thermal properties of mechanochemically pretreated precursors of $BaTiO_3$ synthesis", *J. Therm. Anal. Cal.*, 59 (2000) 1017-1021.

Barry C. (2007). "Ceramic Material Science and Engineering", Springer, USA.

Barsoum M. W. (2003). "FUNDAMENTALS OF CERAMICS", McGraw, USA.

Baskara, N.; Anil Hule, Chutan Bhongale, Ramaswamy Murugan & Hua Chang, "Phase transition studies of ceramic $BaTiO_3$ using thermo-Raman and dielectric constant measurement", *J. Appl. Phys.*, 91 (2002) 10038-10043.

Berbecaru, C.; H. V. Alexandru, C. Porosnicu, A. Velea, A. Ioachim, L. Nedelcu & M. Toacsan, "Ceramic materials $Ba_{(1-x)}Sr_xTiO_3$ for electronics — Synthesis and characterization", Thin *Solid Films*, 2008, 516, (22), 8210–8214.

Blomqvist, M.; Sergey Khartsev & Alex Grishin "Electrooptic ferroelectric $Na_{0.5}K_{0.5}NbO_3$ films", *IEEE Photonics Technology Letters*, 17 (2005) 1638-1640.

Boland, S. W.; Suresh C. Pillai, Weing-Duo Yang & Sossina M. Haile, "Preparation of $(Pb,Ba)TiO_3$ powders and highly oriented thin films by a sol-gel process", *J. Mater. Res.*, 19 (2004) 1492-1498.

Burfoot. Jack C. "*Ferroelectrics*", D. Van Nostrand Company LTD, London.

Buscaglia, M. T.; Massimo Viviani, Vincenzo Buscaglia, Liliana Mitoseriu, Andrea Testino, Paolo Nanni, Zhe Zhao, Mats Nygren, Catalin Harnagea, Daniele Piazza, & Carmen Galassi, "High dielectric constant and frozen macroscopic polarization in dense nanocrystalline $BaTiO_3$ ceramics", *Phys. Rev. B*, 73 (2006) 064114.

Busch, G. & P. Scherrer, *Naturwiss*, 23 (1935) 737.

Chaisan W.; S. Ananta & T. Tunkasiri, "Synthesis of barium titanate-lead zirconate solid solution by a modified mixed-oxide synthetic route", *Curr. Appl. Phys.*, 4 (2004) 182-185.

Chen, Chin-Yen & Hur-Lon Lin, "Piezoelectric properties of $Pb(Mn_{1/3}Nb_{2/3})O_3$-$PbZrO_3$-$PbTiO_3$ ceramics with sintering aid of $2CaO$-FeO_2 compound", *Ceram. Int.*, 30 (2004) 2075-2079.

Cheng, H.; Jiming Ma & Zhenguo Zhao, "Hydrothermal synthesis of PbO-TiO_2 solid solution", *Chem. Mater.*, 6 (1994) 1033-1040.

Cho, Y.; Satoshi Kazuta & Kaori Matsuura, "Scanning nonlinear dielectric with nanometer resolution", *Appl. Phys. Lett.*, 75 (1999) 2833-2834.

Chou, Jung-Fang, Ming-Hong Lin & Hong-Yang Lu, "Ferroelectric domains in pressureless-sintered barium titanate", *Acta Mater.*, 48(2000) 3569.

Deb, K. K.; M. D. Hill & J. F. Nelly, "Pyroelectric characteristics of modified barium titanate ceramics", *J. Mater. Res.*, 7 (1992) 3296-3304.

Deng, X.; Xiaohui Wang, Hai Wen, Liangliang Chen, Lei Chen, & Longtu Li, "Ferroelectric properties of nanocrystalline barium titanate ceramics", *Appl. Phys. Lett.*, 88 (2006) 252905.

DiDomenico, M.; Jr., S. P. S. Porto, S.H. Wemple & R. P. Barman, "Raman spectrum of single-domain $BaTiO_3$", *Phys. Rev.*, 174 (1968) 522-530.

Ding, Z.; R. L. Frost & J. T. Kloprogge, "Thermal Activation of Cooper Carbonate", *J. Mater. Sci. Lett.*, 21 (2002) 981-983.

Eng, L. M., H.; –J. Güntherodt, G. A. Schneider, U. Köpe & J. Muñoz Saldaña, "Nanoscale reconstruction of surface crystallography from three-dimensional polarization distribution in ferroelectric barium-titanate ceramics", *Appl. Phys. Lett.*, 74 (1999) 233-235.

Eng, L. M.; H. –J. Güntherodt, G. Rosenman, A. Skilar, M. Oron, M Katz and D. Eger, "Nondestructive imaging & characterization of ferroelectric domains in periodically poled cristal", *J. Appl. Phys.*, 83 (1998) 5973-5977.

Feteira, A.; Derek C. Sinclair, Ian M. Reaney, Yoshitaka Somiya, & Michael T. Lanagan, "$BaTiO_3$-Based Ceramics for Tunable Microwave Applications", *J. Am. Ceram. Soc.*, 87 (2004) 1082-1087.

Frey, M. H. & D. A. Payne, "Grain-size effect on structure and phase transformations for barium titanate ", *Phys. Rev. B*, 54 (1996) 3158-3167.

Frey, M. H. & D. A. Payne, "Grain-size effect on structure and phase transformations for barium titanate", *Phys. Rev. B*, 54 (1996) 3158-3168.

Ganpule, C. S., V. Nagarajan, B. K. Hill, A. L. Roytburd, E. D. Williams & R. Armes, "Imaging three-dimensional polarization in epitaxial polydomain ferroelectric thin films", J. *Appl. Phys.*, 91 (2002) 1477-1481.

Gao, L.; Yanqiu Huang, Yan Hu & Hongyan Du, "Dielectric and ferroelectric properties of $(1-x)BaTiO_3-xBi_{0.5}Na_{0.5}TiO_3$ ceramics", *Ceram. Int.*, 33 (2007) 1041-1046.

Gopalan, V. & Terence E. Mitchell, "Wall velocities, switching times, and the stabilization mechanism of 180° domains in congruent $LiTaO_3$ crystals", *J. Appl. Phys.*, 83 (1998) 941-954.

Gururaja T. R; Walter A. Schulze, Leslie E. Cross, Robert E. Newnham, Bertram A. Auld & Yuzhong J. Wang, "Piezoelectric Composite Materials for Ultrasonic Transducer Applications", *IEEE*, 32 (1985) 481-498.

Hammer, J. M. "Digital electro-optic grating deflector and modulator", *Appl. Phys. Lett.*, 18 (1971) 147-149.

Harnagea, C.; Alain Pignolet, Marin Alexe & Dietrich Hesse, "Higher-Order Electrochemical Response of Thin Films by Contact Resonant Piezoresponse Force Microscopy", *IEEE*, 53 (2006) 2309-2321.

Hidaka T.; T. Maruyama, M Saitoh & N. Mikoshiba, "Formation and observation of 50 nm polarized domains in $PbZr_{1-x}Ti_xO_3$ thin film using scanning probe microscope", *Appl. Phys. Lett.*, 68 (1996) 2358-2359.

Ianculescu, A.; A. Brăileanu & Georgeta Voicu, "Synthesis, microstructure and dielectric properties of antimony-doped strontium titanate ceramics", *J. Eur. Ceram. Soc.*, 27 (2007) 1123-1127.

Judd, M. D. & M. I. Pope, "Energy of activation for the decomposition of the alkaline-earth carbonates from thermogravimetric data", *J. Therm. Anal.*, 4 (1972) 31-38.

Kamalasanan, M. N.; N. Deepal Kumar & Subhas Chandra, "Structural and microstructural evolution of barium titanate thin films deposited by the sol-gel process", *J. Appl. Phys.*, 76 (1994) 4603-4609.

Kingery. (1960). "Introduction to Ceramics", Second Ed., John Wiley & Sons, USA.

Kingon, A.; "Is the ultimate memory in sight?", *Nat. Mater.*, 5 (2006) 251-252.

Kingon, Angus I,; Jon-Paul Maria & S. K. Streiffer, "Alternative dielectrics to silicon dioxide for memory and logic devices", *Nature*, 406 (2000) 1032-1038.

Kinoshita Kyoichi & Akihiko Yamaji, "Grain-size Effects On Dielectric Properties in Barium Titanate Ceramics", *J. Appl. Phys.*, 47 (1976) 371-373.

Kohlstedt, H.; Y. Mustafa, A. Gerber, A. Petraru, M. Fitsilis, R. Meyer, U. Böttger & R Waser, "Current status and chellenges of ferroelectric memory devices", *Microelectronic Engineering*, 80 (2005) 296-304.

Kong, L. B.; J. Ma, H. Huang, R.F. Zhang & W.X. Que, "Barium titanate derived from mechanochemically activated powders" *J. Alloys Compd.*, 337 (2002) 226-230.

Kotecki, D. E.; "$(Ba,Sr)TiO_3$ dielectrics for future stacked capacitors DRAM", *IBM J. Res. Develop.*, 43 (1999) 367-380.

Kovala, V.; Carlos Alemany, Jaroslav Briancin, Helena Brunckova & Karol Saksl, "Effect of PMN modification on structure and electrical response of xPMN-(1-x)PZT ceramic systems", *J. Eur. Ceram. Soc.*, 23 (2003) 1157-1166.

Kugel, V. D.; G Rosenman & D. Shur, "Electron emission from LiNbO$_3$ crystals with domains of inverted polarization", *J. Phys. D: Appl. Phys.*, 28 (1995) 2360-2364.

L'vov, Boris V. & Valery L. Ugolkov, "Peculiarities of CaCO$_3$, SrCO$_3$ and BaCO$_3$ decomposition in CO$_2$ as a proof of their primary dissociative evaporation", *Thermochim. Acta*, 410 (2004) 47-55.

Laurell, F.; M. G. Roelofs, W. Bindloss, H. Hsiung, A Suna and J. D. Bierlein, "Detection of ferroelectric domain reversal in KTiOPO$_4$", *J. Appl. Phys.*, 71 (1992) 4664-4670.

Lin, Ming-Hong; Jung-Fang Chou & Hong-Yang Lu, "Grain-Growth Inhibition in Na2O-Doped TiO$_2$-Excess Barium Titanate Ceramic", *J. Am. Ceram. Soc.*, 83 (2000) 2155-2162.

Liou, Jih_Wei & d Bi-Shiou Chiou, "Effect of Direct-Current Biasing on the Dielectric Properties of Barium Strontium Titanate", *J. Am. Ceram. Soc.*, 80 (1997) 3093-3099.

Liou, Jing-Kai; Ming-Hong Lin & Hong-Yang Lu, "Crystallographic Facetting in Sintered Barium Titanate", *J. Am. Ceram. Soc.*, 85 (2002) 2931-2937.

Lu, Chung Hsin; Wei-Hsing Tuan & Buh-Kuan Fang, "Effects of Pre-sintering Heat Treatment on the Microstructure of Barium Titanate", *J. Mater. Sci. Lett.*, 15 (1996) 43-45.

Masui, S.; Shunsuke Fueki, Koichi Masuntani, Amane Inoue, Toshiyuki Teramoto, Tetsuo Suzuki & Shoichiro Kawashima, "The Application of FeRAM to Future Information Technology World", *Topics Appl. Phys.*, 93 (2004) 271-284.

Merz Walter J. "Double Hysteresis Loop of BaTiO$_3$ at the Curie Point", *Phys. Rev.*, 91 (1953) 513-514.

Merz, Walter J. "Domain formation and Domain Wall Motions in Ferroelectric BaTiO$_3$ Single Crystal", *Phys. Rev.*, 95 (1954) 690-698.

Meschke, F.; A. Kolleck & G. A. Schneider, "R-curve behaviour of BaTiO$_3$ due to stress-induced ferroelastic domain switching", *J. Eur. Ceram. Soc.*, 17 (1997) 1143-1149.

Millar C. A. "Hysteresis Loss and Dielectric Constant in Barium Titanate", *Brit. J. Appl. Phys.*, 18 (1967) 1689-1697.

Padmaja, G.; Ashok K. Batra, James R. Curie, Mohan D. Aggarwal, Mohammad A. Alim & Ravindra B. Lal "Pyroelectric ceramics for infrared detection applications", *Mater. Lett.*, 60 (2006) 1937-1942.

Park, Yong-I, "Effect of composition on ferroelectric properties of sol-gel derived lead bismuth titanate (PbBi$_4$Ti$_4$O$_{15}$) thi films", *J. Mater. Sci.*, 36 (2001) 1261-1269.

Perry, C. H. & D. B. Hall, "Temperature Dependence of the Raman Spectrum of the BaTiO$_3$", *Phys. Rev. Lett.*, 15 (1965) 700-702

Piticescu, R. M.; P. Vilarnho, L. M. Popescu, R. R. Piticescu, "Hydrothermal synthesis of perovskite based materials for microelectronic applications", *J. Optoelectron. Adv. Mater.*, 8 (2006) 543-547.

Pradeep, P. P.; Subhash H. & Risbud, "Low-temperature synthesis and processing of electronic materials in the BaO-TiO$_2$ system", *J. Mater. Sci.*, 25 (1990) 1169-1183.

Rabe, U.; M. Kopycinska, S. Hirsekorn, J. Muñoz Saldaña, G. A. Schneider & W. Arnold, "High-resolution characterization of piezoelectric ceramics by ultrasonic scanning force microscopy techniques", *J. Phys. D: Appl. Phys.*, 35 (2002) 2621-2635.

Radheshyam, R. & Seema Sharma, "Structural and dielectric propierties of Sb-doped PLZT ceramics", *Ceram. Int.*, 30 (2004) 1295-1299.

Ravez, J. "Ferroelectricity in Solid State Chemistry", *Chem.*, 3 (2000) 267-283.

Razak, K. A.; A. Asadov & W. Gao, "Properties of BST prepared by high temperature hydrothermal process", *Ceram. Int.*, 33 (2007) 1495-1502.

Relva, B. C. (2004). "Ceramic Materials for Electronics", Marcel Dekker, USA.

Roberto, J. "High Dielectric Constant Oxides", *Eur. Phys. J. Appl. Phys.*, 28 (2004) 265–291.

Rosenman, G.; A. Skliar & I. Lareah, "Observation of ferroelectric domain structures by secondary-electron microscopy in as-grown $KTiOPO_4$ crystals", *Phys. Rev. B*, 54 (1996) 6222-6226.

Rousseau, D. L. & S. P. S. Porto, "Auger-like Resonant Interference in Raman Scattering From One and Two-Phonon States of $BaTiO_3$", *Phys. Rev. Lett.*, 20 (1968) 1354-1357.

Rupprecht, G. & R. O. Bell, "Microwave Losses in Strontium Titanate above the Phase Transition", *Phys. Rev.*, 125 (1962) 1915-1920.

S. Maitra, N. Chakrabarty & J. Pramanik, "Decomposition kinetics of alkaline heart carbonates by integral approximation method", *Ceramica*, 54 (2008) 268-272.

Sakabe, Y.; N. Wada & Y. Hamaji, "Grain Size Effects on Dielectric Properties and Crystal Structure of Fine-grained $BaTiO_3$ Ceramics", *J. Korean Phys. Soc.*, 32 (1998) 260-264.

Sato, H. & Kohji Toda, "An Application of $Pb(Zr,Ti)O_3$ Ceramic to Opto-Electronic Devices", *Appl. Phys.*, 13 (1977) 25-28.

Saurenbach, F. & B. D. Terris, "Imaging of ferroelectric domain walls by force microscopy", *Appl. Phys. Lett.*, 56 (1990) 1703-1705.

Sawyer, B. & C. H. Tower, "Salt rochelle as a dielectric", *Phys. Rev.*, 35 (1930) 269-273.

Shao, S.; Jialiang Zhang, Zong Zhang, Peng Zheng, Minglei Zhao, Jichao Li1 & Chunlei Wang, "High piezoelectric properties and domain configuration in $BaTiO_3$ ceramics obtained through the solid-state reaction route", *J. Phys. D: Appl. Phys.*, 41 (2008) 125408 (5pp).

Shepard, R. "Dielectric and Piezoelectric Properties of Barium Titanate", *Phys. Rev.*, 71 (1947) 890-895.

Shiratori, Y.; C. Pithan, J. Dornseiffer & R. Waser, "Raman scattering studies on nanocrystalline $BaTiO_3$ Part II – consolidated polycrystalline ceramics", *J. Raman Spectrosc.*, (2007).

Srivastava, N. & G.J. Weng, "The influence of a compressive stress on the nonlinear response of ferroelectric crystals", *Int. J. Plas.*, 23 (2007) 1860-1873.

Stanford, A. L. "Dielectric resonance in Ferroelectric Titanates in the Microwave Region", *Phys. Rev.*, 124 (1961) 408-410.

Takashi Teranishi, Takuya Oshina, Hiroaki Takeda & Takaaki Tsurimi, "Polarization behavior in diffuse phase transition of $Ba_xSr_{1-x}TiO_3$ ceramics", *J. Appl. Phys.*, 105 (2009) 054111.

Takeuchi, T.; Claudio Capiglia, NaliniBalakrishnan, Yasuo Takeda & Hiroyuki Kageyama, "Preparation of fine-grained BaTiO$_3$ ceramics by spark plasma sintering", *J. Mater. Res.*, 17 (2002) 575-581.

Uchino, Kenji. (2000). "Ferroelectric Devices", Marcel Dekker, USA.

Valasek, J. "Piezo-Electric And Allied Phenomena In Rochelle Salt", *Phys. Rev.*, 17 (1921) 475-481.

Venkateswaran, UD.; Naik VM & Naik R, "High-pressure Raman studies of polycrystalline BaTiO$_3$ ", *Phys. Rev. B*, 58 (1998) 14256.

Vittayakorn, N.; Theerachai Bongkarn & Gobwute Rujijanagul, "Phase transition, mechanical, dielectric and piezoelectric properties of Perovskite (Pb$_{1-x}$Ba$_x$)ZrO$_3$ ceramics", *Physica B*, 387 (2007) 415-420.

Vold, R. E.; R. Biederman, G. A. Rossettu JR., & A. Sacco JR., "Hydrothermal synthesis of lead doped barium titanate", *J. Mater. Sci.*, 36 (2001) 2019-2026.

Wainer, E. & A. N. Salomon, Titanium Alloy Manufacturing Company Elect. Rep. 8 (1942), 9 and 10 (1943).

Whatmore, R. W.; Qi Zhang, Christopher P. Shaw, Robert A. Dorey & Jeffery R. Alcock, "Pyroelectric ceramics and thin films for applications in uncooled infra-red sensor arrays", *Phys. Scr. T.*, 129 (2007) 6-11.

Wilk, G. D.; R. M. Wallace & J. M. Anthony, "High-k Gate Dielectrics: Current Status and Materials Properties Considerations", *Appl. Phys. Rev.*, 89 (2001) 5243-5275.

Wittborn. J.; C. Canalias, K. V. Rao, R. Clemens, H. Karlsson & F. Laurell, "Nanoscale imaging of domain and domain walls in periodically poled ferroelectrics using atomic force microscopy", *Appl. Phys. Lett.*, 80 (2002) 1622-1624.

Wodecka-Dus, B.; A. Lisinska-Cekaj. T. Orkisz, M. Adamczyk, K. Osinska, L. Kozielski & D. Czekaj, "The sol-gel synthesis of barium strontium titanate ceramics", *Mater. Sci. Poland*, 25 (2007) 791-799.

Xiao, C. J.; W. W. Zhang, Z. H. Chi, F. Y. Li, S. M. Feng, C. Q. Jin, X. H. Wang, L. T. Li, & R. Z. Chen, "Ferroelectric BaTiO$_3$ nanoceramics prepared by a three-step high-pressure sintering method", *Phys. Stat. Sol.*, 204 (2007) 874-880.

Xu, H. & Lian Gao, "Hydrothermal sinthesis of high-purity BaTiO$_3$ powders: control of powder and size, sintering density, and dielectric properties", *Mater. Lett.*, 58 (2004) 1582-1586.

Yamashita, Kouchi Harada, Yasuharu Hosono, Shinya Natsume & Noboru Ichinose, "Effects of B-site Ions on the Electrochemical Coupling Factors of (Pb(B'B'')O$_3$-PbTiO$_3$ Piezoelectric Materials", *Jpn. J. Appl. Phys.*, 37 (1998) 5288-5291.

Yamashita, Y.; Yasuharu Hosono & Noboru Ichinose, "Phase Stability, Dielectrcis and Peizoelectric Properties of the Pb(Sc$_{1/2}$Nb$_{1/2}$)O$_3$-Pb(Zn$_{1/3}$Nb$_{2/3}$)O$_3$-PbTiO$_3$", *Jpn. J. Appl. Phys.*, 36 (1997) 1141-1145.

Yoo, J. H.; W. Gao & K. H. Yoon, "Pyroelectric and dielectric bolometer properties of Sr modified BaTiO$_3$ ceramics", *J. Mater. Sci.*, 34 (1999) 5361-5369.

Yun, S. N.; X. L. Wang & D. L. Xu, "Influence of processing parameters on the structure and properties of barium strontium titanate ceramics", *Mater. Res. Bull.*, 2008, 43, (8-9), 1989-1995.

Zhi, Y.; Ruyan Guo, A.S. Bhalla, "Dielectric polarization and strain behavior of Ba(Ti$_{0.92}$Zr$_{0.08}$)O$_3$ single crystal", *Mater. Lett.*, 57 (2002) 349-354.

Zhong, Z. & Patrick K. Gallagher, "Combustion synthesis and characterization of BaTiO$_3$", *J. Mater. Res.*, 10 (1995) 945-952.

Zhu, M.; Lei Hou, Yudong Hou, Jingbing Liu, Hao Wang & Hui Yan, "Lead-free (K$_{0.5}$Bi$_{0.5}$)TiO$_3$ powders and ceramics prepared by sol-gel method", *Mater. Chem. Phys.*, 99 (2006) 329-332.

Part 2

Superconducting Ceramics

4

Sintering Process and Its Mechanism of MgB$_2$ Superconductors

Zongqing Ma and Yongchang Liu

Tianjin Key Lab of Composite and Functional Materials,
School of Materials Science & Engineering,
Tianjin University, Tianjin,
P R China

1. Introduction

1.1 The phase formation mechanism of MgB$_2$ during sintering

The superconductivity at 39 K discovered in MgB$_2$ among simple binary chemical composition attracted much interest in its fabrication techniques and practical applications [1]. MgB$_2$ superconductor exhibits many impressive properties such as highest critical temperature amongst intermetallic superconductors which means low cooling costs, impressive grain boundary transparency to the flow of current which leads to greater critical current density [2-4], comparatively large coherence length which allows a better Josephson junction fabrication, low material cost which will lead to a cheaper superconductor technology, simple crystal structure, etc. Hence, MgB$_2$ superconductors, especially the MgB$_2$ wires and coils, have the outstanding potential to be integrated into diverse commercial applications, such as, magnetic resonance imaging (MRI) [5, 6] , fault current limiters (FCL), Josephson junctions and SQUID [7, 8, 9], transformers, motors, generators, adiabatic demagnetization refrigerators, magnetic separators, magnetic levitation applications, energy storage, and high energy physical applications. But the MgB$_2$ itself is mechanically hard and brittle and therefore not amenable to drawing into the desired wire and tape geometry. Thus, the powder-in-tube (PIT) technique that was used to make the Y-Ba-Cu-O oxide superconductor has been employed in the fabrication of MgB$_2$ wires and tapes these years [10-14]. So far, in-situ sintering, including the in-situ PIT, from the mixture of magnesium and boron is the major method to fabricate MgB$_2$ superconductors (bulks, wires and tapes). The corresponding sintering parameters have a significant influence on the superconducting properties of MgB$_2$. Thus it is necessary to investigate the sintering mechanism of MgB$_2$ superconductors.

The reaction process and MgB$_2$ phase formation mechanism during the sintering have been studied by different methods, such as differential thermal analysis (DTA) [15-21], *in-situ* XRD measurement [22-25], *in-situ* resistance measurement [26, 27] and temperature dependent magnetization (M–T) measurements [28].

1.2 Sintering of Mg-B precursor powders over a wide temperature range

It can be seen from the DTA data of the Mg + 2B$_{amorphous}$ precursor composition shown in Fig. 1 that the first exothermic peak occurs in the temperature range below 650 °C (the melting point of Mg) for samples heated at either 20 °C /min or 5 °C /min [29]. Previous studies have suggested different origins of this peak; some speculate that it is due to the reaction between Mg and impurity B_2O_3 in the original B powder [30, 31], whereas others suggest that it is associated with the solid-solid reaction between Mg and B [16, 18]. In general, there is consensus about the origin of the second and third DTA peaks, which are due to melting of Mg and the liquid-solid reaction between and B, respectively.

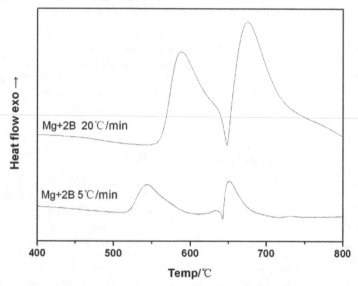

Fig. 1. The measured thermal analysis curves during the sintering of Mg + 2B sample with heating rates of 20 °C·min⁻¹ and 5 °C·min⁻¹[29].

With aim of clarifying the origin of first exothermic peak in the DTA curves, the phase evolution of Mg + 2B$_{amorphous}$ system was detected by in-situ X-ray diffractometer (XRD) during the sintering up to 750 °C and the measured results are shown in Fig. 2. It is found that obvious MgB$_2$ phase peaks can be recognized only after 550 °C. In fact, all the measured results of in-situ resistance, *in-situ* XRD and the temperature dependent magnetization during sintering of a mixed powder of Mg : B = 1 : 2 indicate that the MgB$_2$ phase begins to form before the Mg melting [22, 24, 25, 27, 28]. In this case, the exothermic peak in the DTA curves before the Mg melting occurs, and should be attributed to the solid-solid reaction between Mg and B. The phase formation of MgB$_2$ during the sintering process, therefore, proceeds via solid-solid reaction, Mg melting and, finally, liquid-solid reaction.

Observing the in-situ XRD patterns in Fig. 2 carefully, a lot of Mg is still present as primary phase in the Mg+2B sample sintered at 600 °C. The result implies that the MgB$_2$ phase formed at the solid-solid reaction stage is limited due to the low atomic diffusion rate in the solid state. On the other hand, also as shown in Fig. 2, MgB$_2$ phase forms on a much larger scale and becomes primary phase immediately following completion of the melting of Mg when the sintering temperature reaches 650 °C. As a result, in order to obtain complete MgB$_2$ phase rapidly, the MgB$_2$ superconductors were generally synthesized by sintering at high temperature in the past decade.

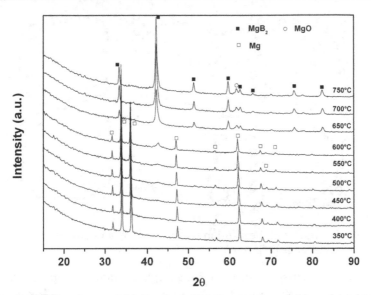

Fig. 2. The in-situ XRD patterns of (a) Cu-doped MgB$_2$ sample and (b) undoped MgB$_2$ sample.

1.3 Sintering process and mechanism of MgB$_2$ superconductors at high temperature

Since most of MgB$_2$ superconductors are prepared by sintering at high temperature, it makes sense to investigate their phase formation process and sintering mechanism at high sintering temperature. At high temperature, the liquid-solid reaction between Mg and B is activated following completion of the melting of Mg. The MgB$_2$ phase formation mechanism at this stage should be very different to that at the solid-solid reaction stage due to the presence of the Mg melt.

A large number of small MgB$_2$ grains exist in the bulk material after the solid-solid reaction, together with residual Mg and B particles. When the sintering temperature is above 650 °C, residual Mg melted and the flowing liquid phase (Mg) increased the diffusion rate of atom and enlarged the contact area of reactants, which leads to a strong and complete reaction between residual Mg and B.

According to our previous study [17], this solid-liquid reaction stage follows Ostwald ripening mechanism and includes three important processes [32]:

i.   *rearrangement of particles.* The molten Mg helps individual particles to slip, spin and reassemble;

ii. *solid-liquid reaction.* The residual B particles are entrapped by the molten Mg, which promotes a strong instantaneous contact reaction;

iii. *solution-reprecipitation and grain growth.* Small MgB$_2$ grains generally have a higher solubility in the liquid phase than larger grains [33] and will dissolve first in the Mg melt to yield an over-saturated solution. This will lead to the precipitation of MgB$_2$ on the surface of existing MgB$_2$ grains, which will lead to further grain growth, as shown in Fig. 3.

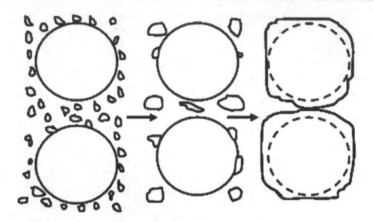

Fig. 3. Schematic illustration of the solution-reprecipitiation and growth process of grains during the liquid-solid stage [17].

According to above discussion, when the sintering temperature rise to 750 °C, the ending point of solid-liquid reaction in the DTA curve (see Fig. 1), complete MgB_2 phase can be obtained (see Fig. 2) and most of MgB_2 grains tend to be regular hexagon in the sample (see Fig. 4, the SEM image of sample sintered at 750 °C).

It is very difficult to calculate the kinetic parameters exactly from the DTA analysis data due to the overlap between the Mg melting and liquid-solid reaction thermal peaks. Hence, only limited kinetic information calculated from the *in-situ* X-ray diffraction measurement has been reported for the liquid-solid reaction between Mg and B. DeFouw *et al.* Ref. [23] reports that the liquid-solid reaction between Mg and B under isothermal conditions can be described by diffusion-controlled models of a reacting sphere with kinetics characterized by diffusion coefficients that increase with temperature from 2×10^{17} to 3×10^{16} s^{-1}, with associated activation energies of $123 \sim 143$ kJ·mol^{-1}. However, a very high heating rate was used in these studies to prevent the reaction between Mg and B occurring below a certain temperature (above the melting point of Mg). As a result, the sintering environment might be quite different from that in traditional sintering methods.

Previous studies on the phase formation mechanism of MgB_2 during liquid-solid sintering between Mg and B are deficient, and further investigation is necessary in addition to advanced test methods.

1.4 Conventional solid-state sintering of MgB_2 superconductors at low temperature

Though high-temperature sintering is the most popular method of synthesizing MgB_2 superconductors till now, the high volatility and tendency of Mg to oxidize at high temperature pose significant challenges to the fabrication of MgB_2 superconductors that exhibit excellent superconductive properties since these processes tend to generate voids and MgO impurities during in-situ sintering. Thus, recent studies have addressed the low-temperature preparation of MgB_2 superconductors in an attempt to reduce the oxidation and volatility of Mg.

As discussed above, the formation of the MgB₂ phase begins at a temperature below the melting point of Mg, which offers the prospect of sintering MgB₂ superconductors at relatively low temperature (i.e. below the melting point of Mg) in an attempt to avoid problems associated with the strong volatility and oxidation of Mg at high temperature. Rogado *et al.* [34] initially fabricated superconducting bulk MgB₂ samples by conventional solid state sintering at a temperature as low as 550 °C (see Fig. 5). This process required a sintering time of 16 hours to form the complete bulk MgB₂ phase, and the samples exhibited inferior superconducting properties than those sintered at high temperature. However, the result indicates that it is possible to fabricate MgB₂ superconductors at low temperature and this work resulted in increased attempts world-wide to develop a low-temperature sintering process for both undoped and doped MgB₂ bulk superconductors.

Fig. 4. SEM image of MgB₂ sample sintered at 750 ℃.

Yamamoto *et al.* [35] found that MgB₂ bulk superconductors prepared by solid state sintering at 600°C for 240 h exhibited improved critical current densities at 20 K (see Fig. 11). This study confirmed the potential of the low temperature sintering technique for the fabrication of bulk MgB₂ superconductors. It also established that poor crystallinity is found to enhance H_{c2}, H_{irr} and J_c in MgB₂ at high fields, whereas strong grain connectivity, reduced MgO impurity content and a smaller grain size increases J_c at low fields.

MgB₂ wires and tapes with improved H_{irr} and J_c can also be prepared by low temperature sintering by an *in-situ* PIT technique. Goldacker *et al.* [36] reports the synthesis of thin, steel-reinforced MgB₂ wires with very high transport current densities at only 640 °C. These authors suggest that the low-temperature annealing could lead to a fine grain structure and a superconducting percolation path with very high associated critical current density. Moreover, the observation of a dramatically-reduced reaction layer between the filament and sheath in their samples is very promising for the production of filaments with small diameters in mono- and multifilamentary wire.

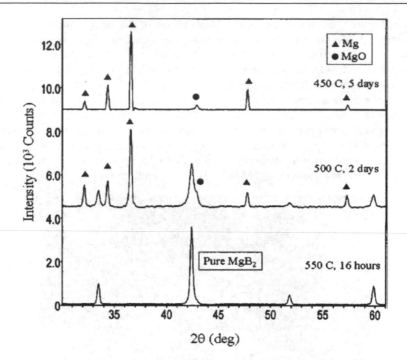

Fig. 5. Powder x-ray diffraction patterns of MgB$_2$ samples prepared using different heating conditions. Markers are placed above the peaks corresponding to Mg and MgO impurities. The unmarked peaks correspond to MgB$_2$ (from Ref. [34]).

Recently, a new process, called two-step heat-treatment, has been developed to fabricate undoped MgB$_2$ bulk superconductors [37]. In this process, short high-temperature sintering at 1100 °C is followed by low-temperature annealing. Samples prepared by this method exhibit, uniquely, well-connected small grains with a high level of disorder in the MgB$_2$ phase, which yields an in-field J_c of nearly one order of magnitude higher than for the samples prepared by single-step sintering at high or low temperature. However, the applicability of the two-step heat-treatment to the fabrication of MgB$_2$ wires has yet to be investigated.

To summarize, these MgB$_2$ superconductors synthesized at low temperature were generally cleaner and denser than the same samples sintered at high temperature due to the reduced volatility and oxidation of Mg, which could improve the connectivity between MgB$_2$ grains. Moreover, sintering at low temperature can also obtain the refined MgB$_2$ grains, which obviously strengthens the grain-boundary pinning. Both of factors are bound to result in the improvement of critical current density in the low-temperature sintered samples compared to the typical high-temperature sintered samples. From this point of view, the low-temperature synthesis might be the most promise and effective method in obtaining the higher J_c in MgB$_2$ superconductors. Hence, it makes sense to clarify the phase formation mechanism of MgB$_2$ during the low-temperature sintering.

1.5 Sintering kinetics of MgB₂ superconductors at low temperature

Analysis of the kinetics of the sintering process can be performed based on the DTA data using different computational methods. Yan *et al.* [18] calculated the activation energy of the solid-solid reaction at low sintering temperature using the Ozawa–Flynn–Wall and Kissinger as 58.2 and 72.8 kJ·mol⁻¹, respectively. The value of the pre-exponential factor calculated using the Kissinger method is 2.0×10^{15} s⁻¹. They also report that the activation energy increases parabolic as the reaction progresses [18]. However, the study by Shi *et al.* [19] draws different conclusions to those of Yan et al. These authors use a new kinetic analysis (based on a variant on the Flynn-Wall-Ozawa method) under non-isothermal conditions and suggest that the solid-solid reaction between Mg and B powders follows an instantaneous nuclei growth (Avrami–Erofeev equation, $n = 2$) mechanism. The values of E decrease from 175.4 to 160.4 kJ·mol⁻¹ with the increase of the conversion degrees (α) from 0.1 to 0.8 in this model. However, the activation energy (E) increases to 222.7 kJ·mol⁻¹ [19] again as the conversion degree reaches 0.9.

On the other hand, the solid-solid reaction between Mg and B exothermal peak is partly overlapped with the Mg melting endothermic peak in the DTA curves. This phenomenon makes it difficult to calculate the kinetics parameters exactly from the thermal analysis data and also could be the reason why the previous results are different from different research groups. It is necessary to further investigate the phase formation mechanism of MgB₂ during the low-temperature sintering combined with advanced test methods.

In our recent work [38], in-situ X-ray diffraction technique is used to measure the degree of reaction between Mg and B as a function of time at several certain temperatures below Mg melting, respectively. Based on these isothermal data, the kinetics analysis of MgB₂ phase formation during the low-temperature sintering is carried out.

Bulk samples of MgB₂ were prepared by a solid-state sintering method using amorphous boron powder (99% purity, 25µm in size), magnesium powder (99.5% purity, 100 µm in size). Several reaction temperatures in the range of 550~600 ℃, below the melting point of Mg, were chosen as the isothermal holding temperatures. Then the samples were fast-heated to the chosen temperature with a rate of 50 ℃/min in order to prevent significant reaction between Mg and B before arriving at the isothermal annealing temperature. The x-ray diffraction measurement started as soon as the sample temperature reached the certain isothermal temperature and it will detect the sample every 15 min till the reaction is over. The weight fraction of synthesized MgB₂ which corresponds to the degree of reaction was calculated from the XRD data of sample obtained after different soaking time according to the External Standard Method.

Fig. 6 illustrates the typical X-ray diffraction patterns of the Mg-B sample isothermally annealed at 575 ℃ for different periods. One can see that no organized MgB₂ peak can be observed when the temperature just reaches 575 ℃. It implies that the reaction between Mg and B did not occur during the rapid heating to the final isothermal holding temperature. As the holding time prolonging, the MgB₂ phase appears and increases gradually while the Mg phase decreases. However, the increase in the intensity of MgB₂ peaks becomes very slow and even stops when it reaches a certain degree despite of longer holding time (ie. longer than 480 min). Similar behavior is also found in the isothermal annealing experiments at 550 ℃ and 600 ℃ (the XRD patterns in not shown here).

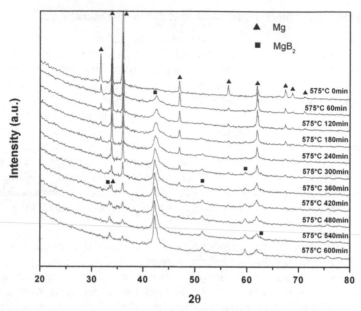

Fig. 6. Typical X-ray diffraction patterns of the Mg-B powder specimen isothermally annealed at 575 ºC for different periods [38].

According to the XRD data, the weight fraction of the synthesized MgB_2 phase, which is considered as the degree of reaction, is calculated using the External Standard Method. Plots of degree of reaction vs. time are given in Fig. 7 for isothermal sintering at 550 ºC, 575 ºC

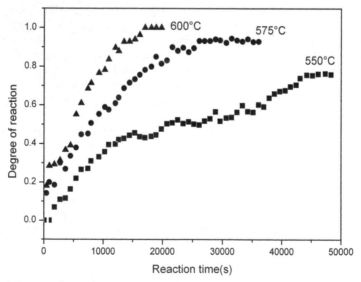

Fig. 7. Plots of degree of reaction vs. time isothermally-annealed Mg-B powder specimens at 550 ºC, 575 ºC and 600 ºC, respectively [38].

and 600 °C, respectively. At 550 °C and 575 °C, the reaction seems stop even though the degree of reaction is still below 100% and there is residual Mg and B in the sample (see Fig. 6). It means that at the final stage of isothermal heating in present work, the reaction rate is so slow that it is difficult to observe the increase in the degree of reaction. At each isothermal annealing temperature, it can be also found that the reaction rate becomes slower and slower with the reaction time increasing and the degree of reaction seems unchanged at last.

Based on these isothermal data, kinetics analysis of the MgB$_2$ phase formation during the low-temperature sintering is carried out. Early kinetics studies employed the currently-accepted kinetic equation:

$$\frac{d\alpha}{dt} = k(T)f(\alpha) \tag{1}$$

Where t represents time, a is the degree of reaction, T is the temperature, k(T) is the temperature-dependent rate constant and $f(a)$ is a function that represents the reaction model. k(T) can be described as:

$$k(T) = A\exp(-\frac{E}{RT}) \tag{2}$$

Where A is the pre-exponential factor, E is the activation energy and R is the gas constant.

Integrating Eq. (1), it comes:

$$g(\alpha) \equiv \int_0^\alpha \left[f(\alpha)\right]^{-1} d\alpha = k(T)t \tag{3}$$

Where $g(a)$ is the integral form of $f(a)$.

After substitution for k(T) and rearranging, it yields:

$$\ln(t) = \frac{E}{RT} + \ln\left[\frac{g(\alpha)}{A}\right] \tag{4}$$

It needs different reaction time to reach certain degree of reaction at different isothermal holding temperatures. According to Eq. (4), at certain degree of reaction, one can plot the lnt~$1/T$ and then do a linear fit. Following this way, the activation energy (E) can be obtained without referring to the reaction modes. Table 1 illustrates the reaction time and activation energy at different degree of reaction of the Mg-B sample. It is recognized that the activation energy (E) firstly decreases when a changes from 0.20 to 0.40 and then increases again when a changes from 0.50 to 0.80. It is indicated that the reaction between Mg and B at low temperature is not controlled by only one kinetics reaction model. At different stage of reaction, the kinetics model is varied.

In order to determine the kinetic modes of the reaction between Mg and B during the low-temperature sintering, Model fitting method is performed. Following this method, The determination of the $g(a)$ term is achieved by fitting various reaction models to experimental data. As described below in Eq. (5), the relationship between $g(a)$ and t should be linear.

$$g(\alpha) = A\exp(-\frac{E}{RT})t \qquad (5)$$

Set of alternate reaction models is linearly-fitted to the experimental data obtained from the is-situ X-ray diffraction results at 575 °C and then obtained results are collected in Table 2. The coefficient r of the contracting cylinder, contracting sphere and one-dimensional diffusion models are the better ones. According the Eq. (5), the intercept t during the linear fitting should be zero. But the intercept of contracting cylinder model is 0.0821, which is too high compared to 0. And the corresponding value of r is also not as good as the case in the Contracting sphere model. Hence, the contracting cylinder model is ignored here. On the other hand, the coefficient r of the contracting sphere is better than that of the one-dimensional diffusion model. But the intercept of one-dimensional diffusion model is more near to 0 than that of the contracting sphere model.

Degree of reaction (a)	Reaction time (t) at 550 °C (s)	Reaction time (t) at 575 °C (s)	Reaction time (t) at 600 °C (s)	Activation energy (E) (kJ/mol)
0.20	5200	1500	520	275.39
0.30	6700	3600	2250	130.58
0.40	11700	5700	4500	114.75
0.50	21000	8100	5100	169.71
0.60	36000	11700	6150	181.71
0.70	42300	14100	7800	202.63
0.80	46800	18900	10260	211.74

Table 1. The reaction time (t) and activation energy (E) at different degree of reaction (α) during the low-temperature sintering of the Mg-B powders [38].

Reaction model	G(a)	Slope (b)	r-square	Intercept (a)
One-dimensional diffusion	α^2	3.49×10⁻⁵	0.9851	-0.0165
Two-dimensional diffusion	$[1-(1-\alpha)^{1/2}]^{1/2}$	2.14×10⁻⁵	0.9568	0.3411
Mampel (first order)	$-\ln(1-\alpha)$	9.20×10⁻⁵	0.9720	-0.0060
Avrami-Erofeev	$[-\ln(1-\alpha)]^{1/2}$	4.53×10⁻⁵	0.9830	0.4432
Avrami-Erofeev	$[-\ln(1-\alpha)]^{1/3}$	3.06×10⁻⁵	0.9730	0.6088
Avrami-Erofeev	$[-\ln(1-\alpha)]^{1/4}$	2.32×10⁻⁵	0.9648	0.6977
Contracting cylinder	$1-(1-\alpha)^{1/2}$	2.54×10⁻⁵	0.9855	0.0821
Contracting sphere	$1-(1-\alpha)^{1/3}$	2.05×10⁻⁵	0.9869	0.0417
Power law	$\alpha^{3/2}$	3.39×10⁻⁵	0.9779	0.0711
Power law	$\alpha^{1/2}$	2.09×10⁻⁵	0.9042	0.4934
Power law	$\alpha^{1/3}$	1.57×10⁻⁵	0.8821	0.6276
Power law	$\alpha^{1/4}$	1.25×10⁻⁵	0.8699	0.7061

Table 2. Linear fitting results of the experimental data obtained from the in-situ X-ray diffraction results at 575 °C by adopting alternate reaction models [38].

According to Eq. (4), the intercept of plot $lnt \sim 1/T$ ($\ln(\frac{g(\alpha)}{A})$), can be obtained from the linear fitting. At certain degree of reaction, if the $g(a)$ is given, the value of A can be calculated. Then $k(T) = A\exp(-\frac{E}{RT})$ can also be determined at certain temperatures. The calculated value of $k(T)$ should be consistent with the linear fitted slope of plot $g(a) \sim t$ if the given $g(a)$ is the valid reaction model.

Following this method, we verify these two models discussed above at 575°C. At the initial stage (a=0.20) of reaction, for contracting sphere model ($1 - (1-\alpha)^{1/3}$), the calculated $k(T)$ is 2.09×10^{-5}, which is comparable to the corresponding value of slope given in Table 2. In the case of one-dimensional diffusion model (α^2), the calculated $k(T)$ is 1.19×10^{-5}, which is much smaller than the corresponding value of the slope presented in Table 2. At the middle stage ((a=0.50) of reaction, the calculated value of $k(T)$ from the contracting cylinder model is 2.18×10^{-5}, which is still comparable to the corresponding value of slope. For the one-dimensional diffusion model, the calculated $k(T)$ is 2.64×10^{-5}, which is smaller than the corresponding value given in Table 2. However, at the final stage (a=0.80), the calculated $k(T)$ from the one-dimensional diffusion model is 3.62×10^{-5}, which is more consistent with the corresponding value presented in Table 2 than that in the case of the contracting cylinder model (the calculated value is 2.33×10^{-5}).

Hence, one can say that the reaction between Mg and B during low-temperature sintering is firstly mainly controlled by the contacting sphere model, which is a kind of the phase boundary reaction mechanism. As the reaction prolongs, the one-dimensional diffusion model, one kind of diffusion-limited mechanism, gradually becomes dominant.

In our previous study [39], we have investigated the MgB$_2$ phase formation process during the sintering based on the phase identification and microstructure observation. It is found that the sintering 'necks' between individual Mg and B particles occurs at the first stage of the sintering. Then the solution active regions form at the sintering necks. With the isothermal heating prolonging, a few Mg and B atoms in the solution active regions can be activated and react with each other. The activated atoms are limited at this initial stage and thus the reaction rate is slow and mainly determined by the phase boundary reaction mechanism. Meanwhile, an MgB$_2$ phase layer is gradually formed at the sintering necks between Mg and B particles and the Mg atoms have to diffuse through the whole MgB$_2$ phase layer to reach the reaction interface, , as shown in Fig. 8. As the reaction prolonging, the synthesized MgB$_2$ layer becomes thicker and thicker, and it should be more and more difficult for Mg atoms to reach the reaction interface through diffusion. Finally, the reaction rate is controlled by the diffusion-limited mechanism. As a result, the reaction rate becomes slower and slower and it takes a very long time to be totally completed due to the slow diffusion rate of Mg. The corresponding activation energy is also decreased firstly and then increased again during the whole reaction process, as discussed above. It is also the reason why the residual Mg is still present even after holding for 10 h at 575 °C.

Based on above analysis and compared with previous studies [18, 19] in which they propose the reaction is controlled by single mechanism, the varied mechanisms are more valid and more consistent with the actual sintering process. The value of activation energy in present work is also comparable to the calculated value in Shi et al.'s study [19] whereas the models are different from theirs. But in their study, the activation energy is calculated using a model-free method, just as in our work.

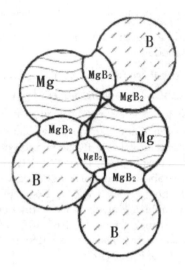

Fig. 8. Schematic illustration of solid-solid reaction between Mg and B particles based on the inter-diffusion mechanism [39].

Based on above discussion, it is concluded that the reaction between Mg and B during the low-temperature sintering is controlled by varied mechanisms. At initial stage, the reaction rate is mainly determined by the phase boundary reaction mechanism. As the reaction prolonging and the synthesized MgB_2 layer increasing, the diffusion-limited mechanism gradually becomes dominant. The corresponding activation energy is also decreased firstly and then increased again.

2. Accelerated sintering of MgB_2 with different metal and alloy additions at low temperature

According to the above sintering mechanism, the reaction between magnesium and boron at low temperature took a very long time to form the complete MgB_2 phase as the result of the low diffusion rate of Mg atom at solid state. In order to rapidly synthesize the complete MgB_2 phase through the low-temperature sintering, improving the diffusion efficiency of Mg atoms to the reaction interface is the key point, especially at the final stage of the sintering.

2.1 The influence of different metals and alloys on the sintering process and superconductive properties of MgB_2

In order to accelerate the diffusion rate of Mg atoms and thus improve sintering efficiency of MgB_2 at low temperature, different metals and alloys dopants were added into Mg-2B sintering system by world-wide research groups.

Shimoyama *et al.* [40] found that a small amount of silver addition decreases dramatically the reaction temperature of magnesium and boron in the formation of bulk MgB$_2$ without degradation of either the critical temperature or the critical current density. Although the added silver forms an Ag-Mg alloy after the heat treatment, these impurity particles exist mainly at the edge of voids in the sample microstructure and therefore do not provide a significant additional restriction to the effective current path. Accordingly MgB$_2$ bulks with excellent J_c properties have been fabricated at a temperature as low as 550 °C with the 3 at.% Ag doping. The sintering time of doped samples is also reduced significantly compared to that required for undoped samples fabricated by low temperature sintering. This effectively widens the processing window for the development of practical, low-cost MgB$_2$ superconductors by reaction at low temperature [40].

Hishinuma *et al.* [41] synthesized Mg$_2$Cu-doped MgB$_2$ wires with improved J_c by sintering at low temperature for 10 h. They found that the formation of the MgB$_2$ phase is improved due directly to the lower melting point of Mg$_2$Cu (568 °C) than Mg (650 °C), which can promote the diffusion of Mg in the partial liquid (the MgB$_2$ phase forms by the diffusion reaction between released Mg from Mg$_2$Cu and amorphous B powder [41]). The J_c of sample prepared in this way can be improved further in Mg$_2$Cu-doped MgB$_2$ wires by sintering at lower temperature (450 °C) for longer time (more than 100 h). The maximum core J_c value of these samples was found to be over 100, 000 A cm^{-2} at 4.2 K in a magnetic field of 5 T for a tape sintered for 200 h. Bulk MgB$_2$ has been fabricated successfully in other studies by Cu-doping and sintering at 575 ºC for only 5 h [42]. Thermal analysis indicates that the Mg-Cu liquid forms through the eutectic reaction between Mg and Cu at about 485 °C, which leads to the accelerated formation of MgB$_2$ phase at low temperature. The SEM images of the sintered Cu-doped samples are shown in Fig. 9. It is observed that the undoped sample is porous and consists of small irregular MgB$_2$ particles and large regular Mg particles which are in poor connection with each other (see Fig. 9a). On the other hand, the MgB$_2$ particles of the doped sample become larger and more regular accompanying with the increasing amount of Cu addition. The doped samples also become denser with the amount of Cu addition increasing for the reason that the MgB$_2$ particles are in better connection with each other and give birth to less voids (see Figs. 9b-9d). Especially, as shown in the Fig. 9d, the MgB$_2$ grains in the (Mg$_{1.1}$B$_2$)$_{0.9}$Cu$_{0.1}$ sample exhibit platy structure with a typical hexagonal shape [42]. The high J_c in MgB$_2$ samples doped with Cu is attributed mainly to the grain boundary pinning mechanism that results from the formation of small MgB$_2$ grains during low temperature sintering. As with the Ag-doped samples, the concentration of Mg-Cu alloy in these samples tends to form at the edge of voids in the microstructure and does not degrade significantly the connectivity between MgB$_2$ grains, which contributes directly to enhanced J_c. Recently, the addition of Sn to the precursor powder has also been observed to assist the formation of the MgB$_2$ phase during low temperature sintering, and bulk Sn-doped MgB$_2$ prepared at 600 °C for 5 h exhibit good values of J_c [43].

Interestingly, although Ag and Sn addition can form a local eutectic liquid with the Mg precursor at lower temperature than the addition of Cu, the Cu has been found to play a more effective role in the improvement of MgB$_2$ phase formation than Ag and Sn at low temperature. The Cu-doped samples take significantly less time to form the primary MgB$_2$ phase than those containing similar concentrations of Ag or Sn at a similar sintering temperature [40, 42, 43]. Grivel *et al.* [44] observed a similar phenomenon in a study of the effects of both Cu and Ag on the kinetics of MgB_2 phase formation. The addition of 3 at.% Cu or Ag to a precursor mixture consisting of Mg and B powders results in a significant increase of MgB_2 phase formation kinetics in the temperature range below the melting point of Mg. The MgB_2 phase forms more rapidly in the precursors containing Cu than those containing Ag. These authors suggest that this behavior might be related to the lower solubility limit of Cu in solid Mg, compared to the case of the Mg-Ag system [44].

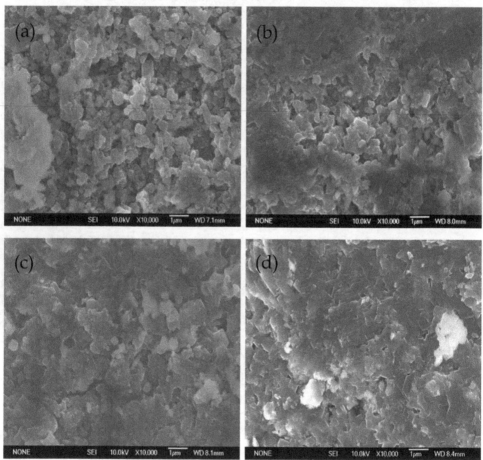

Fig. 9. SEM images of the microstructures of the $(Mg_{1.1}B_2)_{1-x}Cu_x$ samples sintered at 575 °C for 5 h with (a) x = 0.0, (b) x = 0.03, (c) x = 0.05 and (d) x = 0.10, respectively [42].

Based on the above discussion, the assisted sintering of MgB_2 with different metal or metal alloy additions at low temperature is convenient from a practical processing point of view and also reduces the processing time. In addition, these additives tend to be cheap and yield MgB_2 samples with improved J_c. Therefore, this technique appears to be the most promising way of preparing practical, low-cost MgB_2 superconductors at low temperature, compared to the use of different Mg-based precursors and the ball milling pretreatment of precursor powders.

2.2 The mechanism of metal-accelerated sintering at low temperature

The addition of minor metals or metal alloys additions represents the most convenient, effective and inexpensive way of preparing MgB_2 with excellent superconducting properties at low sintering temperatures. As a result, the accelerated sintering mechanism apparent in the effective processing of these samples should be clarified. The accelerated sintering mechanism of precursors containing Cu, for example, has been studied in detail using thermal analysis and activated sintering theory [44, 45, 46].

As discussed in section 1.1, thermal analysis of the sintering process of undoped MgB_2 reveals three peaks corresponding to solid-solid reaction, Mg melting and solid-liquid reaction (see Fig. 10). A similar process was observed in samples of composition $(Mg_{1.1}B_2)_{1-x}Cu_x$ with x = 0.01, 0.03, 0.05 and 0.10, except that the on-set temperature of each the three peaks decreased gradually with increasing amount of Cu addition. It should be noted that an apparent endothermic peak appears at about 485 °C, which is just below the first exothermic peak in the thermal analysis curves of the $(Mg_{1.1}B_2)_{0.90}Cu_{0.10}$ samples. By reference to the binary Mg-Cu phase diagram (see Fig. 11), it is apparent that the Mg-Cu liquid initially forms locally through the eutectic reaction during the sintering process of Mg-Cu-B system, resulting in the appearance of this apparent endothermic peak. The local formation of Mg-Cu liquid in the $(Mg_{1.1}B_2)_{1-x}Cu_x$ samples with x = 0.01, 0.03 and 0.05 is limited for the small amount of Cu added to the precursor, which results in an endothermic signal that is too small to be detected by the thermal analysis measurement in this temperature range [46].

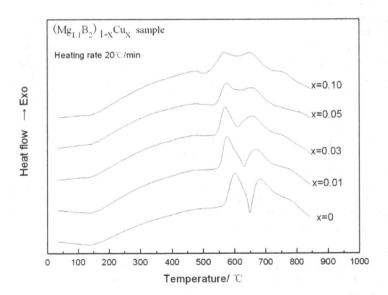

Fig. 10. Measured thermal analysis curves during the sintering of $(Mg_{1.1}B_2)_{1-x}Cu_x$ (with x = 0, 0.01, 0.03, 0.05 and 0.10) samples with an applied heating rate of 20 °C / min [46].

This raises the question of how the presence of a local Mg-Cu liquid accelerates the subsequent formation of MgB_2 phase. (in previous studies [45, 46], the accelerated sintering mechanism of MgB_2 with Cu addition is attributed to the activated sintering).

It is known that activated sintering of MgB_2 by chemical doping facilitates either a lower sintering temperature or a shorter sintering time. German *et al.* [45] have proposed the three criteria for activated sintering systems of solubility, segregation and diffusion as follows:

i.   *Solubility* The additive A must have a high solubility for the base B, while the base B must have a low solubility for the additive A so that the additive can wet the base particles and then exhibit a favorable effect on diffusion.

ii.   *Segregation* During the sintering, the additive must remain segregated at the inter-particle interfaces to remain effective during the entire sintering process.

iii. *Diffusion* The diffusivity of the base metal B in the additive layer must be higher than the self-diffusivity of the base metal B.

Accordingly, an ideal phase diagram for the activated sintering system can be constructed (see Fig. 12). The formation of the MgB_2 phase in the Mg-Cu-B system, is controlled mainly by the diffusion rate of Mg atoms. Only the effect of Cu addition on this diffusion rate is considered to be significant, and the effect of Cu addition on the B atoms can be neglected. Hence, the Mg-Cu-B system can be simplified as an Mg-Cu system for the analysis of the influence of Cu addition on the sintering process. As shown in the Mg-Cu phase diagram (see Fig. 11), Cu addition dramatically decreases the liquidus and the solidus, which implies that the local Mg-Cu liquid can segregate to the interface of Mg particles and therefore meets the *Segregation* criterion in German's study [45]. The high solubility of Cu for Mg (see Figs 11 and 12) enables the Mg-Cu liquid to wet the Mg particles and support the diffusion transport mechanism. As a result, this meets the *solubility* criterion. Generally, the atomic diffusivity in the liquid state is larger than that in the solid state, so the Mg-Cu liquid also meets the *Diffusion* criterion. Collectively, these observations suggest theoretically that the local Mg-Cu liquid meets all the criteria for the diffusion of Mg to B. Cu, therefore, can serve as the activated sintering addition and accelerate the formation of the MgB_2 phase. The conclusion can also be verified by the microstructure observation of the Cu-doped sample sintered at low temperature, as shown in Fig. 13 [46]. It was clear that Cu was concentrated at local region by the edge of voids while Mg was preferentially distributed inside of particles far away from the voids. The result indicated that the Mg-Cu alloys corresponding to the local Cu-Mg liquid during the sintering process mainly concentrated at the edge of voids. Since the void results from the diffusion of Mg atom into B during sintering as mentioned previously [21], the concentration of Mg-Cu alloys at the edge of the voids implied that the Mg-Cu liquid generated and segregated to the interface between Mg particles and B particles at the initial stage of the sintering process and then provided a high transport for the diffusion of Mg into B. After a period of sintering time, Mg was run out and voids formed at the former place of the Mg.

Other metals or metal alloys must first form local liquids with Mg before the formation of MgB_2 phase if they are to serve as activate additives during low temperature sintering. Whether or not these local liquids promote the formation of the MgB_2 phase and activate the sintering mechanism, should be verified by considering the criteria described above (for the case of Cu addition), the ideal phase diagram for the activated sintering system and the binary phase diagram of the added element and Mg. In the section 2.1, the addition of Cu was demonstrated to be more effective than Sn or Ag in accelerating the formation of the

MgB₂ phase at low temperature, even though the latter can form liquid at much lower temperature. Inspection of the relevant binary phase diagram of Sn and Ag with Mg (not shown here) indicates that the solubility limit of Mg in Sn is much lower than that in Cu. Hence, it is more difficult for the Sn-based local liquid to wet the Mg particles and promote the diffusion of Mg according to the *Solubility* criterion. As a result, the activated sintering of MgB₂ with Sn addition is much less efficient than Cu addition. On the other hand, the solubility limit of Mg in Ag is higher than that in Cu and the Ag-based local liquid should wet the Mg particles more easily and accelerate the diffusion of Mg more effectively. However, the solubility limit of Ag in solid Mg is also much higher than for the case of Cu, which means that the amount of local Ag-based liquid present will decrease due to the solution of Ag in the Mg solid. The effect of this is to lower the activated sintering efficiency compared to that obtained for a similar concentration of Cu addition.

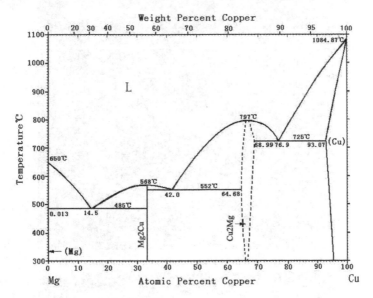

Fig. 11. Binary Mg-Cu phase diagram [48].

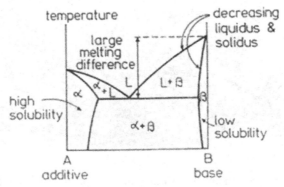

Fig. 12. An ideal phase diagram for the activated sintering system [45].

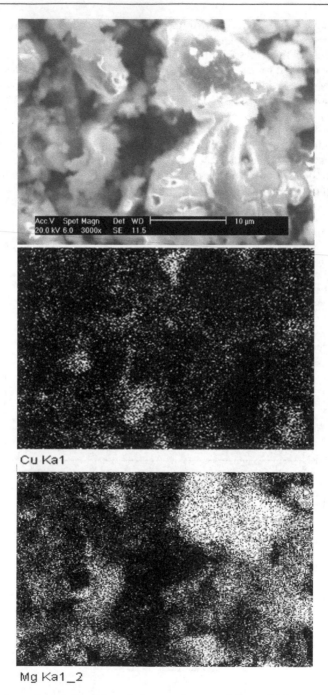

Cu Ka1

Mg Ka1_2

Fig. 13. A secondary electron image and elemental maps of Cu and Mg for $(Mg_{1.1}B_2)_{0.8}Cu_{0.2}$ sample sintered at 575 °C for 5 h [46].

In addition, following these criteria and inspection of the appropriate binary phase diagrams of the additive metal elements with Mg, effective activated addition for the low temperature sintering of MgB_2 could be achieved potentially using lots of candidate metals. Accordingly, Cu was finally determined as effective activator for improving sintering efficiency of MgB_2 in our work.

3. The effect of Cu activator on the microstructure and superconductive properties of MgB_2 prepared by sintering

Cu addition can improve the sintering efficiency of MgB_2 and thus selected as sintering activator. However, whether Cu activator optimizes the microstructure and superconductive properties of MgB_2? To answer this question, the effect of Cu activator on the microstructure of MgB_2 sintered at both low temperature and high temperature were investigated in detail.

3.1 Effect of Cu activator on the reduction of MgO impurity in MgB_2 sintered at high temperature

MgO is always present as the inevitable impurity phase during the sintering process of MgB_2 for the reason that Mg is very reactive with oxygen, which can be supplied by the gaseous O impurity in the protective Ar gas and the oxide impurity (such as B_2O_3 impurity in the B powders) in the starting materials. The presence of MgO impurity may be of great importance and yields a significant effect on the superconductive properties of MgB_2 superconductor. Although the MgO nanoinclusions within MgB_2 grains could serve as strong flux pinning centers when their size were comparable to the coherent length of MgB_2 (.approximately 6~7 nm), the presence of excess MgO phases or largesized MgO particles at the grain boundaries could result in the degradation of grain connectivity [49, 50]. Hence, it is essentially important to control the amount of MgO impurity during the sintering of MgB_2 samples.

In our previous work [51], based on the investigation of the effect of minor Cu addition on the phase formation of MgB_2, it is found that the minor Cu addition (<3 at %) could apparently reduce the amount of MgO impurity in the prepared MgB_2 samples, which provided a new route to govern the oxidation of Mg during the in-situ sintering of MgB_2 samples by altering the amount of Cu addition.

Figure 14 shows the X-ray diffraction patterns of the $(Mg_{1.1}B_2)_{1-x}Cu_x$ (x = 0.0, 0.01, 0.03 and 0.10) samples sintered at 850°C for 30min. It can be seen that all the sintered samples contain MgB_2 as the main phase. In the undoped samples, the MgO peaks are easily recognized, which suggests that some Mg was oxidized during the sintering process and thus MgO was the main impurity in the sintered samples. On the other hand, in the diffracted patterns of the Cu-doped samples, all the MgO phase peaks become weaker and even some peaks identified as MgO phase disappear with the amount of Cu addition increasing. This trend can be observed more clearly from the Fig. 15, which shows the most intense peak (the peak of (200) crystal plane) of MgO in the X-ray diffraction patterns of the sintered $(Mg_{1.1}B_2)_{1-x}Cu_x$ (x = 0.0, 0.01 and 0.03) samples. The results suggest that the minor Cu addition can depress the oxidation of Mg apparently during the in-situ sintering of MgB_2 samples.

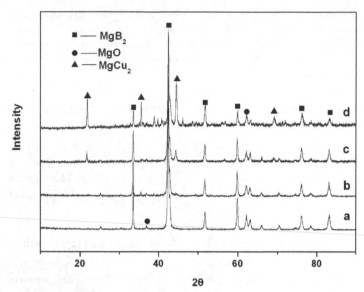

Fig. 14. X-ray diffraction patterns of the prepared $(Mg_{1.1}B_2)_{1-x}Cu_x$ samples sintered at 850°C for 30min with (a) x = 0, (b) x = 0.01, (c) x = 0.03, (d) x=0.10, respectively [51].

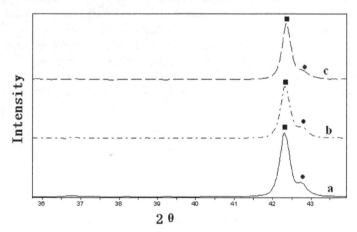

Fig. 15. The enlarged X-ray diffraction patterns around (200) of MgO impurity in the $(Mg_{1.1}B_2)_{1-x}Cu_x$ samples sintered at 850°C for 30min with (a) x = 0, (b) x = 0.01, (c) x = 0.03, respectively [51].

The weight fraction of MgO was calculated from the X-ray diffraction patterns according to the External Standard Method. Fig. 16 shows the weight fraction of MgO versus the amount of Cu addition in the sintered samples. From the figure, it is found that the weight fraction of MgO in the undoped MgB_2 sample is about 16.5%, which is comparable with the previous study [52]. On the other hand, the weight fraction of MgO decreases obviously from 16.5% to 12.5% with the amount of Cu addition increasing from 0.0 to 0.03. However, when the amount of Cu addition increases from 0.03 to 0.10, the weight fraction of MgO almost remains unchanged (see Fig. 16) while the $MgCu_2$ phase increases significantly (see Fig. 14). The excess Cu addition in the $(Mg_{1.1}B_2)_{0.9}Cu_{0.1}$ sample has no significant effect on the further decrease of MgO impurity. Besides, the excess $MgCu_2$ phase in the $(Mg_{1.1}B_2)_{0.9}Cu_{0.1}$ sample (see Fig. 1) may also depress the superconductivity properties of MgB_2 phase. Hence, we conclude that the x=0.03 Cu addition has the best effect on the decrease of MgO impurity during the in-situ sintering of MgB_2 samples.

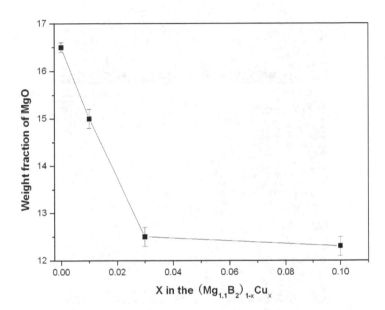

Fig. 16. The weight fraction of MgO versus the amount of Cu addition in the Cu-doped samples sintered at 850 °C for 30 min [51].

The scanning electron microscopy images of the sintered $(Mg_{1.1}B_2)_{1-x}Cu_x$ samples are shown in the Fig. 17. There is an MgO layer on the partial surface of MgB_2 grains in the undoped sample, as shown in Fig. 17a. The MgO layer consists of short MgO whiskers, which is similar with the MgO morphology observed in the study on the oxidation of MgB_2 [53]. On the other hand, in the Cu-doped samples, the amount of MgO impurity decreases with the increasing amount of Cu addition and at the same time the morphology of MgO transits from whiskers to nanoparticles (see Fig. 17b and Fig. 17c). The result of SEM images is consistent with the XRD pattern and they both reveal that the addition of minor Cu can apparently decrease the MgO impurity in the MgB_2 samples. From Fig. 17, it is also found that the MgB_2 grains become larger and more regular accompanying with the increasing amount of Cu addition, which indicates that the Cu addition can also promote the growth of MgB_2 grains at the same time with decreasing the MgO impurity in the prepared MgB_2 samples.

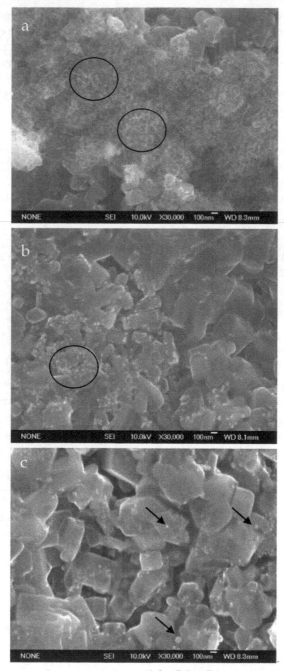

Fig. 17. Scanning electron microscopy images of the $(Mg_{1.1}B_2)_{1-x}Cu_x$ samples sintered at 850 °C for 30 min with (a) x = 0, (b) x = 0.01 and (c) x = 0.03, respectively. The MgO whiskers and nanoparticles are indicated by the black circles and arrows in the figures [51].

In order to investigate the effect of the decreasing MgO impurity induced by the minor Cu addition on the superconductive properties of MgB_2 samples, the corresponding T_c temperatures of all sintered samples were measured. Fig. 18 illustrates the temperature dependence of resistivity for the $(Mg_{1.1}B_2)_{1-x}Cu_x$ (with x = 0, 0.01 and 0.03) samples sintered at 850 °C for 30min. As shown in it, the undoped sample exhibits a slight suppression in the value of T_c compared to the typical pure MgB_2 samples, which can be attributed to the limit of MgB_2 intergranular connection caused by the excessive MgO impurity at the grain boundaries. However, in the Cu-doped samples, the values of T_c are over 38K and slightly increase from 38.1 K to 38.6 K with the increasing amount of Cu addition from x = 0.01 to x = 0.03, which is comparable to the pure MgB_2 samples (39 K). The observation could be explained by the decrease of MgO impurity and the growth of MgB_2 grains resulting from the minor Cu addition.

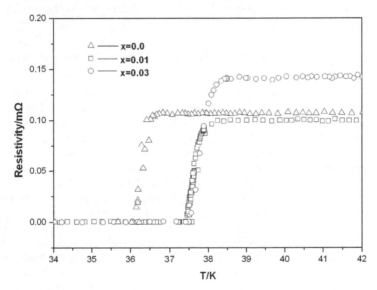

Fig. 18. The temperature dependence of resistivity for the $(Mg_{1.1}B_2)_{1-x}Cu_x$ (x = 0, 0.01 and 0.03) samples sintered at 850 °C for 30min [51].

In summary, the minor Cu addition can decrease the amount of MgO impurity and thus significantly improve the superconductive properties of MgB_2 bulks. However, how the Cu addition avoids the oxidation of Mg during the sintering process of MgB_2 is still to be answered.

It has been indicated that during the sintering process of Mg-Cu-B system, the Mg-Cu liquid locally formed firstly in the presence of Cu through an eutectic reaction. The local Mg-Cu liquid appearing at such low temperature could dissolve some Mg and wrap the neighboring Mg particles, which partly avoided Mg contacting with the gaseous O existing in the interspace of the pressed samples and the oxide impurity (such as B_2O_3 in the B powders) in the starting materials. Hence, the oxidation of Mg during the low-temperature (below the Mg melting point) sintering stage resulting from the gaseous O existing in the interspace and the oxide impurity in the starting materials can be depressed by the presence of local Mg-Cu liquid significantly.

When the temperature was above the Mg melting point (about 650 ºC) during the sintering process, the unreacted Mg after the solid reaction stage would melt and volatilize severely as a result of the high vapor pressure of Mg liquid. The gaseous Mg mixed with the protective Ar gas and could react with the O_2 impurity in the protective Ar gas at such high temperature, which resulted in the increasing amount of MgO impurity deposited in the undoped MgB_2 samples after cooling to room temperature [54], as shown in Fig. 17a. On the other hand, the Cu addition could lower the melting point of Mg at the same time of decreasing the vapor pressure of Mg liquid at high temperature [55]. The decrease of the vapor pressure of Mg liquid led by the Cu addition could reduce the amount of the gaseous Mg from the volatilization of Mg, which thus decrease the amount of MgO impurity resulting from the oxidation of gaseous Mg in the doped samples (as shown in Fig. 17b and 17c).

3.2 The synthesis of lamellar MgB_2 crystalline by Cu activated sintering at low temperature

The microstructure of MgB_2 synthesized by Cu activated sintering at low temperature was also investigated in detail [56]. The SEM images of both sintered Cu-doped sample and undoped sample are illustrated in Fig. 19. Lamellar MgB_2 grains with typical hexagonal shape were observed in the Cu-doped sample sintered at low temperature (see Fig. 19a, denoted by the black arrows). There are few impurities between lamellar MgB_2 grains in the MgB_2 region and the Mg-Cu impurities mainly distribute in the region near the lamellar MgB_2 grains, as shown in Fig. 19a. One can also find that all of the MgB_2 grains in the lamellar structure almost share the same orientation except only a few of them. On the other hand, the MgB_2 grains in the undoped sample sintered at high temperature are nearly in the same size as those in the Cu-doped samples and most of them also exhibit typical hexagonal shape. But their orientation is random, which is the typical characteristic of MgB_2 grains sintered in the traditional solid-state sintering (see Fig. 19b). Hence, the MgB_2 grains in the lamellar structure seem to be in better connectivity with each other and there are also fewer voids between them when compared to the MgB_2 grains in the random orientation, as shown in Fig. 19.

It is proposed that Mg atoms could easily diffuse into B through the path of local Mg-Cu liquid and then react with B forming MgB_2 at the interface between Mg-Cu liquid and B particles. Since the local Mg-Cu liquid only serves as the path for the diffusion of Mg into B and does not react with B until all the Mg is run out, it is always present and provide the constant liquid environment for the nucleation and growth of MgB_2 grains. As we all known, there are two main mechanism forming the lamellar crystalline in the liquid sintering environment, two dimensional nucleation and screw dislocation nucleation. The concentration gradient is the driving force for both mechanisms during the isothermal liquid sintering. The surface of the grains in the lamellar crystalline formed following the two dimensional nucleation are generally more smooth and regular than the screw dislocation [57]. On the other hand, the two dimensional nucleation also needs higher supersaturation than the screw dislocation [58]. In present case, the surface of MgB_2 grains in the lamellar crystalline is smooth and regular and no obvious dislocations and impurities are observed.

Moreover, the supersaturation of MgB$_2$ in the Mg-Cu liquid must be high enough for the two dimensional nucleation due to the comparable less amount of Mg-Cu liquid (the amount of Cu addition is only 8 at%). Hence, it is proposed that the formation mechanism of the present lamellar MgB$_2$ grains is attributed to the two dimensional nucleation. The energy barrier of the two dimensional nucleation could be defined as follows:

$$(C \, / \, Ce)_{crit} = \exp(\pi h \Omega r^2 \, / \, 65 k^2 T^2) \tag{6}$$

$(C \, / \, Ce)_{crit}$ is the critical supersaturation, h is the step height, Ω is the atomic volume, r is the surface energy of crystal, k is the Boltzmann constant and T is the sintering temperature. It can be seen that the energy barrier of the two dimensional nucleation mainly depends

Fig. 19. The SEM images of sintered samples with (a) the Cu-doped sample sintered at 575 °C for 5h and (b) the undoped sample sintered at 750 °C for 1h [56].

on the step height and the surface energy of MgB_2 crystal. On the other hand, only when the MgB_2 supersaturation in the Mg-Cu liquid is higher than the critical supersaturation, the nucleation can start on the surface of B and form the new step. After that, the MgB_2 grains easily grow at this step and form a crystal layer. Accordingly, the schematic of the formation mechanism of the lamellar MgB_2 crytalline is shown in the Fig. 20 with (a) the initial stage, (b) the nucleation and growth stage and (c) the final stage. At the initial stage, the Mg that diffused to the surface of B through the Mg-Cu liquid would react with B as below: Mg + 2B = MgB_2, which can result in the concentration gradient of Mg in the interface (see Fig. 20a). As a result, a lot of Mg could diffuse into the interface and react with B forming MgB_2. Most of produced MgB_2 is dissolved in the Mg-Cu liquid and some MgB_2 will be physically absorbed on the surface of B. When the MgB_2 supersaturation is higher than the critical value for the nucleation, these absorbed MgB_2 will form two dimensional nuclei and produce a new step through the thermodynamic fluctuation. And then the dissolved MgB_2 will deposit on this step and the MgB_2 grains could rapidly grow on this step and form the crystal layer (see Fig. 20b). The new nuclei will continuously forming on the surface of MgB_2 crystal layer and then a new crystal layer will form again. As a result, the lamellar MgB_2 grains are obtained (see Fig. 20c).

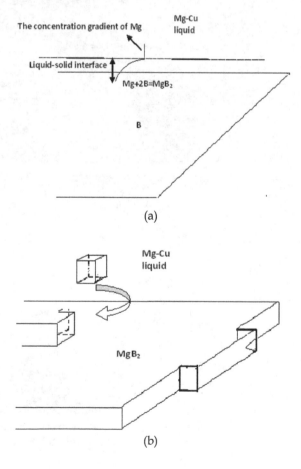

(a)

(b)

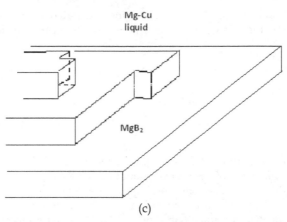

(c)

Fig. 20. The schematic of the formation mechanism of the lamellar MgB$_2$ crytalline with (a) the initial stage, (b) the nucleation and growth stage and (c) the final stage [56].

Fig. 21 shows the temperature dependence of resistivity of the Cu-doped sample and undoped sample. The resistivity of the Cu-doped samples is much lower than the undoped sample in the measured temperature region from 300K to 40K. The resistivity of MgB$_2$ sample should be increased with the addition of Cu, as reported previously [59]. Since the Mg-Cu alloys mainly concentrate around the voids and do not degrade the grain connectivity of MgB$_2$ phase in present sample, the resistivity is ought to keep unchanged and should not be lower than the undoped sample. Hence, the abnormal low resistivity must be attributed to the lamellar structure of MgB$_2$ grains.

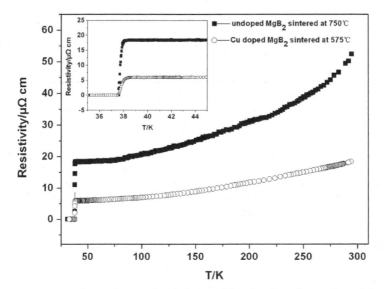

Fig. 21. The temperature dependence of resistivity of the Cu-doped sample and undoped sample [56].

According to the Rowell connectivity analysis, the calculated active cross-sectional area fraction (A_F) represents the connectivity factor between adjacent grains [60]. Here the A_F is estimated as:

$$A_F = \Delta \rho_{ideal} / (\rho_{300K} - \rho_{40K}) \tag{7}$$

$$\Delta \rho_{ideal} = \rho_{ideal(300K)} - \rho_{ideal(40K)} \tag{8}$$

Where ρ_{ideal} is the resistivity of a reference crystal and ρ_T is our measured resistivity at temperature T.

According to the previous studies [13, 60, 61], here the $\Delta \rho_{ideal}$ is 7.3 $\mu\Omega$cm. The results are listed in the Table 3. The value of A_F in the present undoped sample is comparable to that of samples sintered under the similar condition in previous reports [60]. The Cu-doped sample exhibits very excellent A_F, more than two times higher than that of the undoped samples. To further analyze the intergrain connectivity, the residual resistivity ratio (RRR) defined by ρ_{300K} / ρ_{40K} was also estimated, as shown in Table 3. All of the above results indicate that the lamellar MgB$_2$ grains possess much better grain connectivity than the typical morphology of MgB$_2$ grains. It should be pointed out that the onset of the transition temperature ($T_{c(onset)}$) of the Cu-doped sample is higher than that of undoped samples, which might be due to better connectivity and higher crystallinity of the lamellar grains (see Fig. 20a and Fig. 20b). However, the width of transition (ΔT_c) of the Cu-doped sample becomes a little wider than that of undoped sample. The transition broadening can be caused by the small grain size, inhomogeneity, impurities and so on. In present case, the lamellar MgB$_2$ grains shared the same orientation and lead to the intrinsic inhomogeneity in the doped sample, which could be the main factor broadening the transition width.

samples	ΔT_c (K)	$T_{c(onset)}$ (K)	ρ_{40K} ($\mu\Omega$cm)	ρ_{300K} ($\mu\Omega$cm)	RRR	A_P
Undoped MgB$_2$	0.25	38.0	18.388	51.449	2.780	0.221
Cu-doped MgB$_2$	0.40	38.3	5.980	18.355	3.069	0.590

Table 3 The transition temperature ($T_{c(onset)}$), width of transition (ΔT_c), measured resistivity values, residual resistivity ratio (RRR) and active cross-sectional area fraction (A_F) for Cu-doped sample and undoped sample, respectively.

In summary, the lamellar MgB$_2$ grains can be obtained by Cu-activated sintering at low temperature. This lamellar MgB$_2$ grains possess much better grain connectivity than the typical morphology of MgB$_2$ grains synthesized by the traditional solid-state sintering. Together with the proper methods increasing the pinning, the present lamellar MgB$_2$ grains might result in the further improvement of J_c.

4. References

[1] J. Nagamatsu, N. Nakagawa, T. Muranaka, Y. Zentani and J. Akimitsu: Nature, 2001 410 63-64.
[2] D.C. Larbalestier, L.D. Cooley, M.O. Rikel, A.A. Polyanskii, J. Jiang, S. Patnaik, X.Y. Cai, D.M. Feldmann, A. Gurevich, A.A. Squitieri, M.T. Naus, C.B. Eom, E.E. Hellstrom,

R.J. Cava, K.A. Regan, N. Rogado, M.A. Hayward, T. He, J.S. Slusky, P. Khalifah, K. Inumaru and M. Haas: Nature, 2001 410 186-189.

[3] K. Kawano, J.S. Abell, M. Kambara, N. Hari Babu and D.A. Cardwell: Appl. Phys. Lett., 2001 79 2216.

[4] V. Cambel, J. Fedor, D. Gregusova, P. Kovac and I. Husek: Supercond. Sci. Technol., 2005 18 417.

[5] V. Braccini, D. Nardelli, R. Penco and G. Grasso: Physica C, 2007 456 209-217.

[6] A. Serquis, L. Civale, J.Y. Coulter, D.L. Hammon, X.Z. Liao, Y.T. Zhu, D.E. Peterson, F.M. Mueller, V.F. Nesterenko and S.S. Indrakanti: Supercond. Sci. Technol., 2004 17 L35-L37.

[7] D. Mijatovic, A. Brinkman, D. Veldhuis, H. Hilgenkamp, H. Rogalla, G. Rijnders, D.H.A. Blank, A. V. Pogrebnyakov, J.M. Redwing, S.Y. Xu, Q. Li and X.X. Xi: Appl. Phys. Lett, 2005 87 192505.

[8] S.A. Cybart, K. Chen, Y. Cui, Q. Li, X.X. Xi and R.C. Dynes: Appl. Phys. Lett., 2006 88 012509.

[9] K. Chen, Y. Cui, Q. Li, X.X. Xi, S.A. Cybart, R.C. Dynes, X. Weng, E.C. Dickey and J.M. Redwing: Appl. Phys. Lett., 2006 88 222511.

[10] S. Jin, H. Mavoori, C. Bower and R.B. van Dover: Nature, 2001 411 563-565.

[11] S. Jin, R.C. Sherwood, R.B. Vandover, T.H. Tiefel and D.W. Johnson: Appl. Phys. Lett., 1987 51 203-204.

[12] H. Kumakura, A. Matsumoto, H. Fujii and K. Togano: Appl. Phys. Lett., 2001 79 2435-2437.

[13] P.C. Canfield, D.K. Finnemore, S.L. Bud'ko, J.E. Ostenson, G. Lapertot, C.E. Cunningham and C. Petrovic: Phys. Rev. Lett., 2001 86 2423-2426.

[14] A. Serquis, L. Civale, D.L. Hammon, J.Y. Coulter, X.Z. Liao, Y.T. Zhu, D.E. Peterson and F.M. Mueller: Appl. Phys. Lett., 2003 82 1754-1756.

[15] P.C. Canfield, D.K. Finnemore, S.L. Bud'ko, J.E. Ostenson, G. Lapertot, C.E. Cunningham and C. Petrovic: Phys. Rev. Lett., 2001 86 2423-2426.

[16] A.N. Baranov, V.L. Solozhenko, C. Lathe, V.Z. Turkevich and Y.W. Park: Supercond. Sci. Technol., 2003 16 1147-1151.

[17] J.C. Grivel, R. Pinholt, N.H. Andersen, P. Kovac, I. Husek and J. Homeyer: Supercond. Sci. Technol., 2006 19 96-101.

[18] Q.R. Feng, X. Wang, X.Y.Wang and G.C. Xiong: Solid State Commun., 2002 122 459.

[19] Q.R. Feng, X. Chen, Y.H. Wang, X. Wang, G.C. Xiong and Z.X. Gao: Physica C, 2003 386 653.

[20] Q.R. Feng, C. Chen, J. Xu, L.W. Kong, X. Chen, Y.Z. Wang, Y. Zhang and Z.X. Gao: Physica C, 2003 411 41-46.

[21] W. Goldacker, S.I. Schlachter, B. Obst and M. Eisterer: Supercond. Sci. Technol., 2004 17 S490.

[22] M. Bhatia, M.D. Sumption, S. Bohnenstiehl, S.A. Dregia, E.W. Collings, M. Tomsic and M. Rindfleisch: IEEE Trans Appl. Suprcond., 2007 17 2750-2753.

[23] G.H.S. Price: Int. Met., 1938 62 143.

[24] W.Z. Ostwald: Phys. Chem., 1900 34 495.

[25] W. Goldacker, S.I. Schlachter, B. Obst, B. Liu, J. Reiner and S. Zimmer: Supercond. Sci. Technol., 2004 17 S363-S368.

[26] M. Maeda, Y. Zhao, S.X. Dou, Y. Nakayama, T. Kawakami, H. Kobayashi and Y. Kubota: Supercond. Sci. Technol., 2008 21 032004.

[27] S. Soltanian, X.L. Wang, J. Horvat, S.X. Dou, M.D. Sumption, M. Bhatia, E.W. Collings, P. Munroe and M. Tomsic: Supercond. Sci. Technol., 2005 18 658–666.

[28] W.K. Yeoh, J. Horvat, J.H. Kim, X. Xu and S.X. Dou: Appl. Phys. Lett., 2007 90 122502.

[29] Y.C. Liu, Q.Z. Shi, Q. Zhao and Z.Q. Ma: J. Mater. Sci: Mater. Electron., 2007 18 855.

[30] M.S.A. Hossain, J.H. Kim, X. Xu, X.L. Wang, M. Rindfleisch, M. Tomic, M.D. Sumption, E. WCollings and S.X. Dou: Supercond. Sci. Technol., 2007 20 L51–L54.

[31] J.H. Kim, X. Xu, M.S.A. Hossain, D.Q. Shi, Y. Zhao, X.L. Wang, S.X. Dou, S. Choi and T. Kiyoshi: Appl. Phys. Lett., 2008 92 042506.

[32] G.H.S. Price: Int. Met., 1938 62 143.

[33] W.Z. Ostwald: Phys. Chem., 1900 34 495.

[34] N. Rogado, M.A. Hayward, K.A. Regan, Y. Wang, N.P. Ong, H.W. Zandbergen, J.M. Rowell and R.J. Cava: J. Appl. Phys., 2002 91 274-277.

[35] A.Yamamoto, J. Shimoyama, S. Ueda, Y. Katsura, S. Horii and K. Kishio: Supercond. Sci. Technol., 2005 18 116-121.

[36] W. Goldacker, S.I. Schlachter, B. Obst, B. Liu, J. Reiner and S. Zimmer: Supercond. Sci. Technol. 2004 17 S363–S368.

[37] M. Maeda, Y. Zhao, S.X. Dou, Y. Nakayama, T. Kawakami, H. Kobayashi and Y. Kubota: Supercond. Sci. Technol., 2008 21 032004.

[38] Z.Q. Ma and Y.C. Liu: Mater. Chem. Phys., 2011 126 114-117.

[39] Q.Z. Shi, Y.C. Liu, Q. Zhao and Z.Q. Ma: J. Alloys Compd., 2008 458 553-557.

[40] J. Shimoyama, K. Hanafusa, A. Yamamoto, Y. Katsura, S. Horii, K. Kishio and H. Kumakura: Supercond. Sci. Technol., 2007 20 307-311.

[41] Y. Hishinuma, A. Kikuchi, Y. Iijima, Y. Yoshida, T. Takeuchi and A. Nishimura: IEEE Trans. Appl. Supercond., 2007 17 2798-2801.

[42] Z.Q. Ma, Y.C. Liu, Q.Z. Shi, Q. Zhao and Z.M. Gao: Supercond. Sci. Technol., 2008 21 065004.

[43] Z. Q. Ma, H. Jiang and Y. C. Liu: Supercond. Sci. Technol., 2010 23 025005.

[44] J.C. Grivel, A. Abrahamsen and J. Bednarcik: Supercond. Sci. Technol., 2008 21 035006.

[45] C.M. Kipphut and R.M. German: Sci. sintering, 1988 20 31-41.

[46] Z.Q. Ma, Y.C. Liu, Q.Z. Shi, Q. Zhao and Z.M. Gao: Mater. Res. Bull., 2009 44 531-537.

[47] D.K. Finnemore, J.E. Ostenson, S.L. Bud'ko, G. Lapertot and P.C. Canfield: Phys. Rev. Lett., 2001 86 2420-2422.

[48] The Materials Information Society, Binary Alloy Phase Diagram, 2nd edn plus updates, Metals Park, OH: ASM International (1996).

[49] J.H. Kim, S.X. Dou, D.Q. Shi, et al: Supercond. Sci. Technol., 2007 20 1026-1031.

[50] C.H. Jiang, H. Hatakeyama, H. Kumakura: Physica C, 2005 423 45-50.

[51] Z. Q. Ma, Y.C. Liu, Q.Z. Shi, Q. Zhao and Z.M. Gao: J. Alloys Compd. 2009 471 105-108

[52] J.H. Kim, S.X. Dou, J.L. Wang, et al: Supercond. Sci. Technol., 2007 20 448-451.

[53] D. Yang, H. Sun, H. Lu, et al: Supercond. Sci. Technol., 2003, 16 576-581

[54] X.Z. Liao, A.C. Serquis, Y.T. Zhu, et al: Appl. Phys. Lett., 2002 80 4398-4400.

[55] V. Ganesan, H. Feufel, F. Sommer, et al: Metall. Mater. Trans. B, 1998 29 807-813.

[56] Z. Q. Ma, Y.C. Liu and Z.M. Gao: Scripta Mater. 2010 63 399-402

[57] A. E. Nielsen: J. Crystal Growth, 1984 67 289-310.

[58] L. M. Fabietti and R. Trivedi: Metall. Trans. A, 1991 22 1249-1258

[59] K. Singh, R. Mohan, N. Kaur, N. K. Gaur, M. Dixit, V. Shelke and R. K. Singh: Physica C, 2006 450 124–128.

[60] J. M. Rowell, S. Y. Xu, X. H. Zeng, A. V. Pogrebnyakov, Q. Li, X. X. Xi, J. M. Redwing, W. Tian and X. Pan: Appl Phys Lett, 2003 83 102-104.

[61] R. H. T. Wilke, S. L. Bud'ko, P. C. Canfield, D. K. Finnemore, R. J. Suplinskas and S. T. Hannahs: Physica C, 2005 424 1-16.

Part 3

Dielectrics and Opto-Electronic Materials

5

Sintering to Transparency of Polycrystalline Ceramic Materials

Marta Suárez[1,2], Adolfo Fernández[1,2],
Ramón Torrecillas[2] and José L. Menéndez[2]
[1]ITMA Materials Research
[2]Centro de Investigación en Nanomateriales y Nanotecnología (CINN).
Consejo Superior de Investigaciones Científicas (CSIC) –
Universidad de Oviedo (UO) – Principado de Asturias
Spain

1. Introduction

There is currently a high demand for advanced materials for different types of applications (see Fig. 1.1) in which besides a high mechanical performance, a partial or total transparency in a given spectral range is required. Transparent ceramics become more and more important for applications in which materials are subject to extremely high mechanical and thermal stress in combination with optical properties. More recently, interest has focused on the development of transparent armor materials (ceramic) for both military and civil applications. Also, the development of new optoelectronic devices has extended the use of ordinary optical materials to new applications and environments such as temperature (IR) sensor, optical fiber communications, laser interferometers, etc. A considerable fraction of these new devices operates in aggressive environments, such as ovens, radiation chambers and aerospace sensors. In such cases, the sensitive electronic component must be preserved from the extreme external condition by a transparent window. Transparent and coloured ceramics are also often used as wear and scratch resistant parts such as bearings and watch glasses as well as for their aesthetic Properties in synthetic opals and rubies.

In science and technology the word transparent is used for those components that show clear images regardless of the distance between the object and the transparent window. Clear transparency is achieved when after transmission through the window the light does not undergo noticeable absorption or scattering. This applies, for example, to some glasses, single-crystalline and polycrystalline transparent ceramics.

Most of ordinary optical transparent materials, glasses, polymers or alkali hydrides are soft, weak and/or brittle. Figure 1.2 shows the transmittance spectrum of BK7, a commercial glass widely used for visible optics. However, this material presents a strong absorption in the IR range making it of no use for this spectral range. Furthermore, this material shows a very low melting point (559°C) so it cannot be used at high temperature.

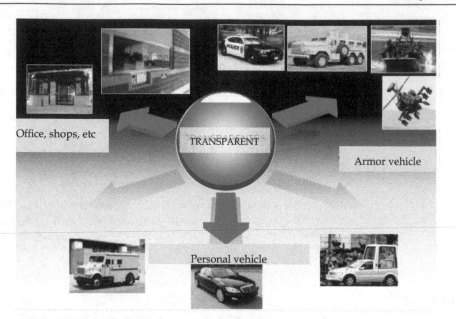

Fig. 1.1. Applications in which transparent materials are required.

However, some ceramic materials, such as corundum, spinel, yttria and YAG, do not show any absorption in a large range between IR and UV (see Fig. 1.2) and are suitable to work under extreme conditions due to their chemical stability and high mechanical performance. Many transparent ceramics are single crystal materials grown from the melt or by flame processes. However, the growth and machining of single crystals is an expensive task, which largely limits their scale up of production and, therefore, their range of applications.

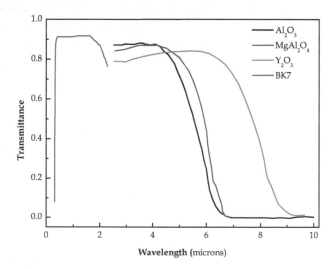

Fig. 1.2. Transmittance spectra of different materials.

Most of these problems can be solved with the use of polycrystalline materials as they show similar mechanical, chemical and thermal stability compared to single crystals and can be produced in complex shapes and are not size limited as single crystals. Among them, only a small number of materials such as MgF_2, ZnS and ZnSe are used for optical applications, but generally their use is limited to coatings and thin films and their mechanical behavior is not appropriate. For this reason, the current trend moves towards the production of polycrystalline transparent ceramics which, by providing more versatility to properties and to the production of complex shapes, will broaden the fields of application. Compared to single crystals, polycrystalline materials have a complex microstructure consisting on grains, grain boundaries, secondary phases and pores. These factors have a great influence on the optical properties, resulting in transparent, translucent or opaque materials, as can be seen in figure 1.3.

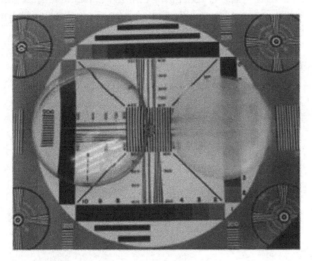

Fig. 1.3. Light scattering on polycrystalline ceramics.

In order to become transparent, the ceramic should be poreless and have optically perfect crystal boundaries and crystals. The surface of a pore is a boundary between phases with sharply different optical properties, and therefore, it intensely reflects and refracts light. A large number of pores makes ceramics opaque. Pores may be intercrystalline or intracrystalline. The elimination of intracrystalline pores, even if they are submicron in size, is a much longer process than elimination of closed intercrystalline pores. Intercrystalline pores occur at crystal boundaries which are sinks of vacancies, and this makes their removal easier. The presence of a second phase on the crystal boundaries which has different optical properties from the main crystalline phase leads to reflection and refraction of light and makes the ceramic less transparent. For this reason, transparent ceramics are obtained from raw material of high purity and the amount of additive is chosen so that they completely dissolve in the solid solution with the main phase.

The crystals in ceramics should be optically perfect, this is the absence of optical defects: pores, inclusions of the second solid phases, aggregate boundaries, and dislocations. In ceramics from optically anisotropic crystals an additional scattering of light arises on the boundaries because of their arbitrary crystallographic orientation.

The amount of light scattering therefore depends on the wavelength of the incident radiation, or light. For example, since visible light has a wavelength scale on the order of hundreds of nanometers, scattering centers will have dimensions on a similar spatial scale. Most ceramic materials, such as alumina and its compounds, are formed from fine powders, yielding a fine grained polycrystalline microstructure which is filled with scattering centers comparable to the wavelength of visible light. Thus, they are generally opaque as opposed to transparent materials. The degree of transparency also depends on the thickness of the component. These losses increase with growing thickness and result in translucent behavior. Transparency, which is independent of the thickness, is only possible for materials with an in-line transmission close to the theoretical maximum. These materials do not cause losses by absorption or scattering. These considerations lead us to the requirements to be observed by manufacturers of transparent ceramics.

2. Ceramic materials

Different materials have been proposed to prepare transparent ceramics such as alumina, spinel, yttria, AlON, YAG, etc (see Fig. 2.1).

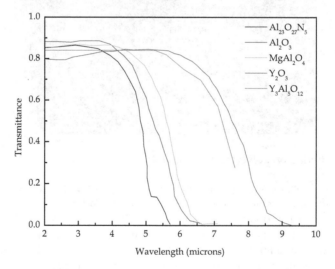

Fig. 2.1. Transmittance spectra of ceramic materials.

Aluminium oxynitride spinel ($Al_{23}O_{27}N_5$, AlON) is a cubic material and one of the leading candidates for transparent armor. This material is very useful for aircraft, missiles, IR and laser windows, etc (Mccauley et al., 2009) as it is stable up to 1200°C, relatively light weight and resistance to damage due to oxidation or radiation. AlON is optically transparent (≥80%) in the near ultraviolet, visible and near infrared regions of the electromagnetic spectrum (see Fig. 2.1). It is four times harder than fused silica glass, 85% as hard as sapphire and nearly 15% harder than magnesium aluminate spinel (table 1). The incorporation of nitrogen into aluminium oxide stabilizes a crystalline spinel phase, which due to its cubic crystal structure and unit cell, is an isotropic material which can be produced as a transparent ceramic nanomaterial. T. M. Hartnett et al (Hartnett et al., 1998) have obtained transparent AlON in the IR region by sintering at 1880°C in a graphite oven. However, this material is not stable under low N_2 pressure and at temperatures higher than 1640°C. Furthermore, this material is susceptible to oxide in air atmosphere at temperatures higher than 1200°C making it not useful as projection lamps (Wei, 2009).

Material	Strength (MPa)	Knoop hardness (Kg/mm2)	Modulus (GPa)	Density (g/ml)	Melting point (K)
AlON	300	1950	323	3.68	2425
Sapphire	700	1500-2200	345-386	3.98	2300
Yttria	150	720	174	5.03	2430
Spinel	100-200	1400	273	3.58	2400
YAG	200-250	1215	300	4.55	1950
ZrO_2	210	1100	180-200	6.52	2128

Table 1. Physical properties of ceramic materials.

Zircona (ZrO_2) is a material with a cubic structure that can be used as a transparent material in different fields: domestic, industrial and military. It is not amongst the hard ceramics (table 1), and its high refractive index near 2.2 is similar to some perowskites. This may explain why few investigations have been published on sintered (polycrystalline) transparent ZrO_2. Single crystals of cubic zirconia are, however, important as artificial gemstones. Their high refractive index gives rise to a high degree of brilliance that comes close to diamonds. In the future the flexibility of powder technology in producing more complex shapes may stimulate the substitution of zirconia single crystals by sintered transparent decorative and optical products. Peuchert et al., (2009) have obtained a transparent material with 65% transmittance at 600 nm using vacuum atmosphere followed by hot isostatic pressing using titanium oxide as sintering aid.

Yttrium oxide (Y_2O_3) is another material with a cubic structure that can be used on pipes for discharge lamps or heat-resistant windows due to its corrosion resistance, thermal stability and transparency in a wide range of the electromagnetic spectrum (figure 2.1). Similarly, when introducing rare earths in its structure, it can be used as the active medium in solid-state lasers. Lukin et al., (1980) have obtained transparent yttria after sintering at 1900°C under vacuum. Anderson (1998) has developed a ceramic Nd:Y_2O_3 by conventional sintering. Greskovish et al., (1973) have developed a laser Y_2O_3:Nd^{3+} doped with ThO_2. La_2O_3 doped Y_2O_3 is of interest for IR applications because it is one of the longest wavelength transmitting oxides. It is a refractory with a melting point of 2430°C and has a moderate thermal expansion coefficient (table 1). A major consideration is the low emissivity of yttria, which limits background radiation upon heating. In particular, lasers with ytterbium as dopant allow the efficient operation both in CW (Kong et al., 2005) and in pulsed regimes (Tokurakawa et al., 2007).

Magnesium aluminate spinel ($MgAl_2O_4$) is a transparent ceramic with a cubic crystal structure. This material shows an excellent optical transmission from 0.2 to 5.5 micrometers and high values of hardness (table 1) making it useful for a wide range of optical applications, including electronic and structural windows in the IR region (see Fig. 2.1). Optical quality transparent spinel has been produced by sinter/HIP, hot pressing, and hot press/HIP operations, and it has been shown that the use of a hot isostatic press can improve its optical and physical properties (Patel et al., 2000). Several authors have worked on obtaining materials of transparent spinel, as the case of Cook et al., (2005). These authors have synthesized spinel powders which subsequently formed by cold isostatic pressing and sintered in a hot pressing machine by applying a mechanical pressure of 4 MPa obtaining a transparent material. Lu et al., (2006) have obtained a transparent spinel material after sintering at low temperature and high pressure (2-5 GPa). Morita et al., (2008) have obtained a transparent spinel material after sintering at 1300°C in plasma sintering equipment (Spark Plasma Sintering, SPS), reaching 47% of transmittance at 550nm. Spinel offers some processing advantages over AlON such as the fact that spinel is capable of being processed at much lower temperatures than AlON, and has been shown to possess superior optical properties within the infrared (IR) region.

Yttrium aluminium garnet ($Y_3Al_5O_{12}$, YAG) is a polycrystalline material with has a cubic structure that can be used in structural (table 1) and functional applications. YAG has been considered a prime candidate for its use as a matrix in oxide-oxide composites in gas turbine engines (Mah et al., 2004). Also, it can be used as substrate for dielectric components, prisms and mirrors, as a part of discharge lamps, high intensity lamps as well as an active medium for the production of lasers (Ikesue et al., 1995), since it is able to accept trivalent cations in its structure, especially rare earths and transition metals. In this sense, YAG can be doped with Yb^{3+} for diode-laser (Takaichi et al., 2003), with Er^{3+} (Qin et al., 2004) widely used in medical applications or Eu^{3+} (Shikao et al., 2001) used as cathode ray tube. Finally, Nd^{3+} doped YAG is one of the most popular laser materials (Savastru et al., 2004). Due to its high thermal conductivity has been widely used in commercial, medical (ophthalmology, cosmetic, dental, etc), military and industry since its discovery in 1964. Three stable phases exist in the Y_2O_3–Al_2O_3 system: an orthorhombic perowskite$YAlO_3$ ("YAP") with an Y/Al ratio 1:1, an alumina-rich cubic garnet $Y_3Al_5O_{12}$ ("YAG"), and an yttria rich monoclinic phase with composition $Y_4Al_2O_9$ ("YAM"). Wen et al., (2004) have obtained a transparent polycrystalline material with a transmittance of 63% in the visible range and 70% in the IR region, using a solid state reaction of alumina and yttria and after sintering at 1750°C in vacuum atmosphere (see Fig. 2.1 for transparency in the IR). Reverse-strike coprecipitation is a common synthesis route for multicationic systems (Li et al., 2000a). Li et al., (2000b) have synthesized a material by the reverse coprecipitation route using aluminum and yttrium nitrates as precursors. The material was sintered at 1700°C in vacuum.

Traditionally, alpha alumina (α-Al_2O_3) has been considered one of the most widely used structural ceramics as basic matrix in many industrial applications, due to its good mechanical properties (table 1), refractory character and its chemical stability in harsh environments. Also, α-Al_2O_3 presents no absorption (see Fig. 2.1) from the near UV (>0.2 μm) until the IR range (> 5 μm). However, due to the birefringent character of alumina, an additional light scattering directly related to the grain size takes place in polycrystalline materials. This light scattering is originated at the boundaries between two alumina grains with different crystalline orientations. Two approaches can be followed to minimize/reduce this scattering: (1) increasing the texture of the polycrystalline material during the forming stage to minimize the misorientations between grains (2) decreasing the grain size as much as possible, which usually implies starting with nanometer sized alumina particles and often making use of non conventional sintering techniques, such as microwave (Jiping Cheng et al., 2002) or spark plasma sintering. However, since Coble's work in the 60's (Coble, 1962), the possibility of using other ceramic materials than the cubic ones for optical applications was demonstrated.

One way to avoid grain growth during sintering and obtain denser materials with reduced porosity is based on the modification of the diffusion mechanisms at high temperature by the introduction of second phases in grain boundaries or by doping with various elements that change the state of the charges at the grain boundary. In the case of alumina many additives have been proposed, mainly metal oxides (NiO, CoO, ZnO, SnO_2, Y_2O_3, MgO, ...) (Budworth, 1970). Magnesia (MgO) (Shuzhi et al., 1999) is one of the most used doping agents in the case of alumina as it inhibits the grain growth and promotes the development of a more uniform grain structure. The role of MgO has been widely studied since the work of Coble in the 60's, indicating that small additions of MgO ($\leq$ 0.25%) allowed obtaining alumina with a density close to theoretical. Bennison and Harmer (1990a) have also studied the role of MgO in the sintering of alumina, highlighting the effect it has on all the parameters that control the sintering of alumina. Handwerker et al. (1989) have also proposed that the addition of magnesia to alumina reduces the chemical heterogeneity due to the presence of impurities, reducing or eliminating the liquid phase generated by the abnormal growth of grains. Jiping Cheng et al., (2002) studied the influence of the presence of MgO on the transparency of alumina and observed that using MgO as sintering aid allows obtaining alumina samples with a more uniform grain size (equiaxed) and no porosity after microwave sintering.

3. Processing and sintering of transparent ceramic materials

Although many transparent ceramics are single crystal materials, transparent polycrystalline ceramics have different advantages such as low price, ease of manufacture, mass-production and more versatility in properties. Furthermore, the capability of producing complex shapes could broaden the fields of application.

The classical issues of ceramic processing are related with powder quality, purity, defect free processing and elimination of minor defects and pores. When the aim is to reach transparency, requirements will be similar from a qualitative point of view but noticeably more restrictive from a quantitative point of view. In addition, microstructural features such as maximum pore and grain size (the latter for birefringent materials) are critical in scattering process. The requirements for these parameters depend on the wavelength at which transparency is desired. In general, as it will be discussed in section 4, it is assumed that defect size must be < λ/10 in order to obtain transparent materials.

Then, in this chapter the processing and sintering techniques most widely used in ceramic manufacturing will be revised, describing the key parameters to be controlled and their effect on the optical properties of the final material.

Ceramic fabrication methods can be classified in different categories depending on the starting materials involved (gas, liquid or solid phase). Polycrystalline ceramics are usually manufactured by compacting powder to a body which is then sintered at high temperatures. The geometry, production volume and characteristic requirements for the component govern the choice of manufacturing process. Alternatively, ceramic materials can be simultaneously formed and sintered when pressure assisted sintering techniques are used (see Fig. 3.1).

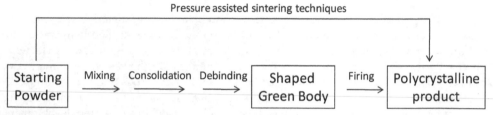

Fig. 3.1. Flow chart for the production of polycrystalline ceramics.

Polycrystalline ceramic processing

The processing steps involved in conventional fabrication of ceramics can be divided in two main parts; formation of green body and firing, and both of them must be carefully controlled in order to avoid residual porosity in the final material.

In the first part, the preparation of shaped green body from ceramic powders, three operations can be identified: mixing, consolidation and debinding. Mixing process includes the preparation of stable slurries from ceramic powders by addition of dispersants or pH controllers and the incorporation of binders and other additives. Consolidation can be done directly from wet slurries or after preparing conditioned powders. Finally, a debinding process for removing the additives used in previous actions before sintering is required.

Although the properties of the starting ceramic powders also play an important role in their behaviour during shaping and sintering, in this review the different ceramic powder synthesis methods will not be described. The description will be focused on the influence of the different processing steps starting from commercial raw materials.

Wet processing of ceramic powders is usually done in order to avoid the formation of undesired powder aggregates that could be responsible of microstructural defects in the final material. As it has been previously mentioned, when the aim is to obtain a transparent material, the critical flaw size above which an important loss of transmittance is observed is very small.

In the processing of ceramics, colloidal suspensions, consisting of a dispersion of solid particles in a liquid, are of particular interest. They are being used increasingly in the consolidation of ceramic powders to produce the green body. Compared to powder consolidation in the dry state, colloidal methods can lead to better packing uniformity in the green body which, in turn, leads to a better microstructural control during firing. Moreover, colloidal suspensions are usually prepared in order to obtain conditioned powders to be consolidated by dye pressing.

Stability of colloidal suspensions depends on particle size and their surface properties. The particles must not be too large otherwise gravity will produce rapid sedimentation. On the other hand, if the attractive force between the particles is large enough, the particles will collide and stick together, leading to rapid sedimentation of particle clusters.

Flocculation will therefore occur unless some process is promoted to produce repulsion between the particles which is sufficiently strong to overcome the attractive force. There are several ways for achieving this, but the most commonly used are:

1. Electrostatic stabilization in which the repulsion between the particles is based on electrostatic charges on the particles
2. Steric stabilization in which the repulsion is produced by uncharged polymer chains adsorbed onto to the particle surfaces
3. Electrosteric stabilization, consisting of a combination of electrostatic and steric repulsion, achieved by the adsorption of charged polymers (polyelectrolytes) onto the particle surfaces.

Rheological measurements are widely used to characterize the properties of colloidal suspensions. They can be used as a method of analysis as, for example, in determining the optimum concentration of dispersant required to stabilize a suspension by measuring the viscosity of the suspension as a function of the concentration of dispersant added.

Once the stable colloidal suspension is prepared, two alternative processes can be distinguished; direct consolidation in order to form a green body or drying under controlled conditions in order to produce a ready to press powder.

In the first case, the processes are known as colloidal forming techniques and their interest is due to the complete deagglomeration of starting powders, reducing the generation of defects. However, other limitations such as difficulties for obtaining larger or complex parts with simultaneously thin and thick cross sections due to density gradients in the green body or problems like the differential sedimentation due to the particle size distribution in the starting powders are found. Slip-casting will be described as an example of colloidal forming technique.

Slip-Casting

Slip casting, with or without pressure, constitutes an ideal combination of dewatering and shaping. As much of the slurry liquid at or near the mould surface is absorbed in the pores in the mould, a layer of solid is formed by the interlocking solid particles in the region near the mould surface. As the process continues, this solid layer increases so long as the mould pores continue to absorb the liquid of the slurry.

The biggest advantage of this process is its versatility in terms of shape, size and materials applicability. It can also accommodate a range of particle sizes, working with typical particle sizes from a fraction of a micron (green densities of ~ 40-50%) to several microns. It can also work fairly well with finer, nanoscale particles. A comparison between slip casting and uniaxial pressing of yttria ceramics show how the minimization of density gradients in green bodies prepared by slip casting allows obtaining more homogeneous materials in terms of translucency and microstructure (J. Mouzon et al., 2008).

However, slip casting has limitations, which, from an operational point of view, are primarily its slow casting rates, with thickness proportional to the square root of casting time, and hence increased cost of casting and of drying to avoid cracking, as well as costs for preparing and maintaining a large mould inventory and facilities for mould storage and drying. Using pressure to accelerate the dewatering process improves the productivity of the process. The main difference in comparison with slip casting is that the water is not removed by capillary suction (negative pressure in the plaster mould) but by pressurisation (positive slip pressure). Control of the filtration process is based on four parameters; the pressure differential on the body, the liquid-medium viscosity, the specific surface area of the slip's solids content and the body porosity (body formation is dependent on the permeability of the layer of body that has already formed from filtered material).

In order to solve those problems a great diversity of advanced forming techniques has been developed. Amongst them, aqueous injection moulding, centrifugal slip casting, direct coagulation casting, electrophoretic casting, gelcasting, hydrolysis assisted solidifications, etc can be mentioned. In all cases, the objective is to obtain a very homogenous green body in order to facilitate the preparation of defect free materials.

However, in other cases, stable suspensions of ceramic powders are prepared in order to produce dry powders especially conditioned for forming by pressing. In this case, the most critical issue is to avoid the formation of hard agglomerates that could lead to formation of defects in the final material which could not be removed during firing. Two drying techniques are especially suitable for this purpose; spray drying and freeze-drying.

Spray-Drying vs. freeze drying

The two main drying techniques for obtaining ready to press powders are spray drying and freeze drying. In spray drying process, the solvent is eliminated by evaporation when the slurry is passed through a chamber at a temperature over the solvent boiling point, whereas during freeze drying the suspension is previously frozen and then water is removed by sublimation. The two processes are similar, except for energy flow. In the case of spray drying, energy is applied to the droplet, forcing evaporation of the medium resulting in both energy and mass transfer through the droplet. In spray freeze drying, energy only is removed from the droplet, forcing the melted to solidify. Both techniques are schematized in figure 3.2.

Spray drying is the most widely used industrial process involving particle formation and drying. It is highly suited for the continuous production of dry solids in either powder, granulate or agglomerate form from liquid feedstocks as solutions, emulsions and pumpable suspensions.

There are three fundamental steps involved in spray drying.

1.   Atomization of a liquid feed into fine droplets.
2.   Mixing of these spray droplets with a heated gas stream, allowing the liquid to evaporate and leave dried solids.
3.   Dried powder is separated from the gas stream and collected.

Spray drying involves the atomization of a liquid feedstock into a spray of droplets and contacting the droplets with hot air in a drying chamber.

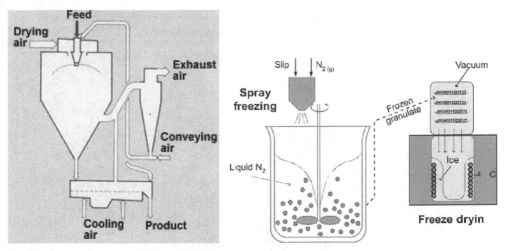

Fig. 3.2. Spray-drying and freeze–drying processes.

The sprays are produced by either rotary (wheel) or nozzle atomizers. Evaporation of moisture from the droplets and formation of dry particles proceed under controlled temperature and airflow conditions. Powder is discharged continuously from the drying chamber. Operating conditions and dryer design are selected according to the drying characteristics of the product and powder specification.

Spray formation is usually coupled to freeze drying process. This technique is named spray freeze drying. This process consists of

1. Atomization of liquid solutions or suspension using ultrasound, one-or two fluid nozzles or vibrating orifice droplet generators
2. Freezing of the droplets in a cryogenic liquid or cryogenic vapour
3. Ice sublimation at low temperature and pressure or alternatively atmospheric freeze-drying using a cold desiccant gas stream

The advantage of using conditioned powder for obtaining transparent ceramic materials is known (I. Amato et al., 1976). The enhanced behaviour during compaction of spray dried or freeze dried powders leads to a more homogeneous distribution of particles in the green body and finally a reduction in the residual porosity of the material. Nevertheless, the incorporation of additives during slurry preparation or binders for favouring soft granulation of powders makes an additional process before firing necessary. This process is named debinding.

Binders, which are used in the slip casting process or in pressing, give the green body a sufficient strength by gluing together particles at their boundary surfaces. Usually those binders are based on polyvinyl alcohols (PVA), polyacrylate or cellulose. High-polymeric compounds such as cellulose and polysaccharides work as plastification agents. They make the flow of ceramic masses during extruding possible.

The thermal treatment of the debinding process destroys the polymers by oxidation or combustion in oxygen containing atmosphere. Very often it is an uncontrolled reaction of high reaction rate inside the shaped part creating a high gas pressure, which can lead to ruptures of the compact. Debinding process is schematized in figure 3.3.

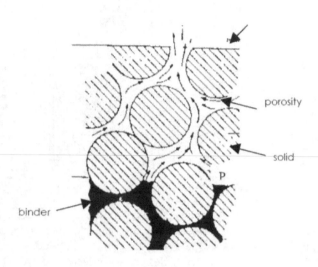

Fig. 3.3. Description of debinding process.

Debinding process is a critical step. In order to overcome the problems in thermal debinding, solvent debinding has been widely adopted by industry. In this process, a portion of the binder can be chemically removed by using solvents like acetone, trichloroethane or heptane. A large amount of open porosities, after solvent debinding, allows the degraded products to diffuse to the surface easily. A more environmentally friendly method is given by binder compositions containing water-soluble components, like polyethylene glycol.

Sintering

Finally, the green body is sintered by heating at high temperature in order to eliminate porosity and obtain the desired microstructure. The driving force for sintering is the reduction in surface free energy of the consolidated mass of particles. This reduction in energy can be accomplished by atom diffusion processes that lead to either densification of the body (by transport matter from inside the grains into the pores) or coarsening of the microstructure (by rearrangement of matter between different parts of the pore surfaces without actually leading to a decrease in the pore volume). The diffusion paths for densification and coarsening are shown in figure 3.4 for an idealized situation of two spherical particles in contact. The densification processes remove material from the grain boundary region leading to shrinkage whereas coarsening processes produce microstructural changes without causing shrinkage.

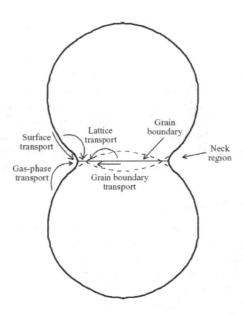

Fig. 3.4. Schematic representation of phenomena involved during firing of ceramics.

When the objective is to obtain transparent materials, it is necessary to control the microstructure evolution during ceramic sintering. The grain growth that occurs during sintering reduces the optical transmittance due to two reasons. Firstly, the pore size distribution in the sintered material is proportional to average grain size. Secondly, in the case of birefringent materials as alumina, the higher the grain size the more important the light scattering.

Grain growth inhibition can be attained by different methods. The most common strategies for reducing grain growth during sintering are the design of complex sintering cycles as two step sintering or the use of dopants for blocking coarsening mechanisms. Two-step sintering has been reported for preparing nanograin ceramics. In the process, the sample is first heated to a higher temperature to achieve an intermediate relative density (above 75%), and then rapidly cooled down and held at a lower temperature until it is fully dense (Z. Chen et al., 2008).

On the other hand, additives can impede the grain growth of ceramics during sintering. In particular, there are many studies treating the alumina grain growth inhibition using different dopants (S.J. Bennison et al., 1990a) (I. Alvarez et.al, 2010). An innovative method for controlling alumina grain growth is described by M. Suárez et al., (2011). In this case, the dopant used is an alumina precursor in order to obtain pure alumina after sintering. The effect of alumina precursor doping on process kinetics and microstructure evolution is discussed.

However, the residual porosity of materials after conventional sintering is usually high enough to reduce noticeably their transmittance or even lead to opaque materials. Thus, an additional process for removing that porosity is usually needed. An effective method for attaining complete densification is the post-HIP treatment. In hot isostatic pressing, the sample predensified, until only closed porosity is present, is placed in a pressure vessel. The pressurizing gas is introduced by means of a compressor to achieve a given initial gas pressure, and the sample is heated to the sintering temperature. During this time the gas pressure increases further to the required value and collapses around the sample, thereby acting to transmit the isostatic pressure to the sample. The use of this technique for improving transparency has been shown by different authors (K. Itatani et al., 2006 and M. Suárez et al., 2010).

Pressure assisted sintering techniques

The simultaneous application of pressure and heat is also used in the Hot-Pressing and Spark Plasma Sintering, methods that are known as pressure assisted sintering techniques. In this case, a mechanical uniaxial pressure is applied to the sample placed in a dye by a vertical piston while the system is heated. Graphite is the most common die material because it is relatively inexpensive, is easily machined, and has excellent creep resistance at high temperatures. The rate of densification can be deduced by following the piston displacement. Typically, in Hot- Pressing, the sintering temperature is chosen to achieve full densification in 30 min to 2 h. Some guidance for selecting the appropriate hot pressing temperature may be obtained from pressure sintering maps, but trial and error is usually done. Pressure is often maintained during the cooldown step.

A sintering method with a configuration similar to Hot-Pressing has been recently developed. It is named Spark Plasma Sintering (SPS) and its main characteristic is that a pulsed DC current is directly passed through the graphite die while uniaxial pressure is applied. The characteristics include (a) high heating rate, (b) the application of a pressure, and (c) the effect of the current. The main advantage in comparison with other sintering techniques is the high heating rates that can be applied during sample sintering. The description of SPS process and a comparison of cycle duration with Hot Pressing are shown in figures 3.5 and 3.6

These features have a great influence on the characteristics of the materials obtained. The extremely short processing times allow obtaining special microstructures in the final material that are unattainable by other sintering techniques. Thus, it is possible to fabricate a dense material with an average grain size similar to starting powders. There are many scientific works showing the potential of this technique for obtaining a wide diversity of transparent polycrystalline materials (B. N. Kim et al., 2007), (C. Wang et al., 2009), (G. Zhang et al., 2009), (R. Chaim et al., 2006). Nowadays, one of the challenges related with this technique is the scaling-up in order to obtain large samples or complex shape components.

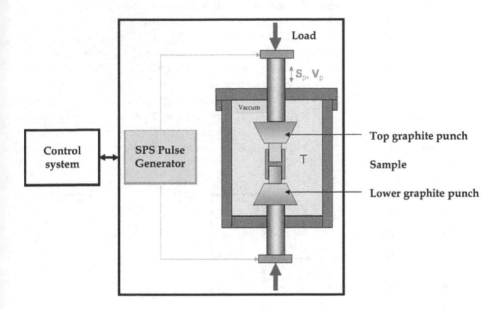

Fig. 3.5. Description of SPS process.

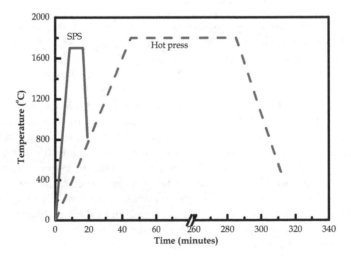

Fig. 3.6. Comparison of cycle duration between Hot-Pressing and Spark Plasma Sintering

4. Characterization of transparent materials

Transparency represents the ability of materials to allow the transmission of light through them. When traversing a transparent material, a light beam propagates along the same direction before and after traversing the material (see Fig. 4.1a). However, perfectly transparent materials are rare in nature and most of them present the so-called scattering centres. When the concentration of scattering centres is sufficiently large, the material still allows the transmission of light, but the transmitted light beam does not only propagate along the incident direction, but diffuse, off-scattered, beams can be detected; in this case the material becomes translucent (see Fig. 4.1b). In this work, we shall refer only to the transparency of a material, which is measured as the real in-line transmittance. When the amount of scattering centres is higher, the materials scatter so much light that they become opaque (see Fig. 4.1c).

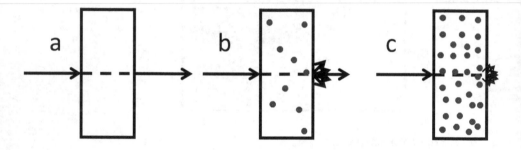

Fig. 4.1. Light propagation through a transparent (a), a translucent (b) and a highly dispersive, almost opaque (c) material.

Therefore, it becomes necessary to define a measurement technique that allows distinguishing between translucency and transparency. In this case, the relevant parameter is the real in line transmittance (Appetz & Bruggen, 2003) or RIT. In this case, a collimated light beam impinges on the sample and the detector is placed far away, usually 1 m, from the sample. This way, for ordinary detectors and light beams, the light scattered > 0.5° will not be detected. Any non-absorbing material must be spatially homogeneous with respect to its dielectric properties in order to become transparent. From an electromagnetic point of view, a defect, or scattering centre, is a spatial region in which a difference in the diagonal dielectric constant (refractive index) is present. The effect of dielectric heterogeneities is manifested through light scattering phenomena, which leads to losses both in the optical quality of the materials and in the total transmitted energy. In ceramics, the main scattering sources are given by the presence of pores, second phase inclusions, rough interfaces... (see Fig. 4.2). Whereas the roughness at interfaces can be minimized by an adequate polishing of the surfaces and the presence of second phases can be also made negligible by selecting pure materials, the presence of pores and the effects of grain boundaries can only be minimized by an adequate processing and sintering. In most cases, and particularly in cubic materials such as yttria, YAG, AlON, spinel, pores are by far the main source of scattering and most of the efforts in the literature have been devoted to minimization of this kind of scattering. In non-cubic materials, of which alumina is the most intensively studied material, there is an additional scattering given by the dependence of the value of the refractive index with the orientation of the crystalline grain, which we shall refer to hereafter as scattering due to grains.

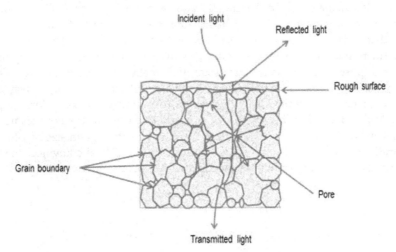

Fig. 4.2. Light scattering sources in polycrystalline ceramics.

Whenever a light beam reaches a surface separating two media with different refractive indices (see Figure 4.3), it deviates according to the Snell's law:

$$n_1 \sin\theta_1 = n_2 \sin\theta_2 \tag{1}$$

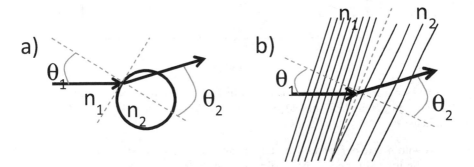

Fig. 4.3. Light refraction at (a) a matrix-pore boundary and (b) between two grains with different crystallographic orientations in a birefringent polycrystalline material.

A model based on geometrical optics was first produced by A. L. Dalisa and R. J. Seymour in which a spread function describing the effect on an incident collimated beam of randomly oriented spatial regions with a difference in the refractive index between zero and $\pm \Delta n_{max}$

was developed for ferroelectric ceramics (Dalisa & Seymour, 1973). This model was further applied by R. Appetz et al. to study the off-specular transmittance in polycrystalline alumina (Appetz & Bruggen, 2003), a birefringent material for which n_1 can be taken as 1.768 and n_2 as 1.76. Considering a grain boundary placed at 45°, the deviation of the light beam at this interface is calculated to be around 0.26°. If a beam light traverses a 1 mm thick polycrystalline alumina sample, in which the grain size is around 0.5 μm, the number of refractions that a light beam undergoes is not below several thousands. This gives an idea of the large scattering undergone by the beam. This discussion also illustrates why the larger the number of pores (for the same pore size), the smaller the real in line transmittance. However, following the same reasoning, one would expect the real in line transmittance to be higher for increasing grain size, whereas the experimental observations (Appetz & Bruggen, 2003) indicate that the amount of light scattered out of the normal incidence is larger in a sample with a grain size of 20 μm than in one with a grain size of 1 μm (Figure 4.4). This indicates that the geometrical model used above is not valid anymore to study the scattering of objects with dimensions of the order of the wavelength of the light.

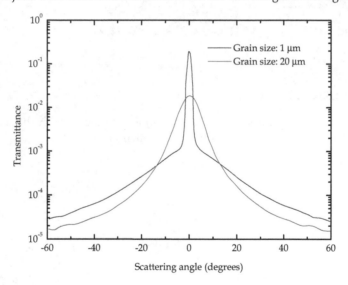

Fig. 4.4. Angular scattering profiles corresponding to 0.8 mm thick polycrystalline alumina samples with 1 μm (solid line) and 20 μm (dotted line) grain size (taken from Appetz & Bruggen, 2003).

In systems with a few defects and by using the Mie equation (Mie, 1908), it is possible to determine the intensity and directionality of the scattered radiation. This formulation is derived from the Maxwell equations and it is, therefore, valid for arbitrary defect sizes. However, as the Mie equation is complex to handle, it becomes helpful to use approximations. Two approximations are often used (Bohren & Huffmann, 2010; van de Hulst, 1957): Rayleigh-Gans-Debye scattering for large particles: $2\pi r >> \lambda$ and Rayleigh scattering for small particles: $2\pi r << \lambda$, with r being the radius of the scattering centre. When dealing with transparent materials or materials close to transparency, the number of pores and their dimensions verifies the conditions for the second approach (Rayleigh scattering) and, therefore, we'll focus our analysis under that simplification. According to this well-known approach, the intensity of electromagnetic radiation with wavelength λ scattered by an object of diameter d and refractive index n, at a distance R and angle of scattering θ is given by:

$$I = I_0 \frac{1+\cos^2\theta}{2R^2} \left(\frac{2\pi}{\lambda}\right)^4 \left(\frac{n^2-1}{n^2+2}\right)^2 \left(\frac{d}{2}\right)^6 \tag{2}$$

The Rayleigh scattering cross section for a single particle can be expressed as:

$$\sigma_s = \frac{2\pi^5}{3} \frac{d^6}{\lambda^4} \left(\frac{n^2-1}{n^2+2}\right)^2 \tag{3}$$

In case a group of N scattering particles is considered, the scattering cross section is given by N times the single scattering cross section. This implies that the scattering due to pores of diameter d in a ceramic is much larger, by a factor of 2^6, than that due to pores of diameter $d/2$. If the size of the pores is reduced to half the initial size, the individual scattering is reduced by a factor of 2^6, whereas the number of scattering centres must be increased by a factor 2^3 if the total porosity is kept constant. This implies that, in total, by reducing the pore size by a factor of 2, the total dispersion is reduced by a factor 2^3, almost an order of magnitude. Therefore, in order to improve the real in line transmittance, two approaches can be followed: the porosity is strongly reduced or, as shown above, the size of the pores is kept much smaller than the wavelength of the radiation used.

Since the pioneering works by Peelen and Metselaar (Peelen and Metselaar, 1974), lots of efforts have been devoted to analyse and model the effects of the different scattering sources on the real in line transmittance. It is well known that the real in line transmittance decays exponentially with thickness, d, according to:

$$RIT = (1 - R_s)^2 \cdot \exp(-\gamma d) \tag{4}$$

where γ, the total scattering, is the sum of the scattering coefficients due to grains and pores: $\gamma = \gamma_{gb} + \gamma_p$. On the other hand, R_s accounts for the total reflected light at the air-material and material-air boundaries. It is well known that the amount of light reflected from a surface between two media (Born & Wolf, 2005) with refractive indices n_1 and n_2 at normal incidence is given by $R_s=(n_1-n_2)^2/(n_1+n_2)^2$. In this case, one of the media is air ($n_1=1$) and the other medium is the material under study ($n_2=n$). R. Appetz and M. P. B. van Bruggen calculated the light scattering coefficients due to pores and different grain orientations in non-cubic materials, considering just one pore size and one grain size (Appetz & Bruggen, 2003). The expressions given for those coefficients are:

$$\gamma_{gb} = \frac{3r_{gb}\pi^2}{\lambda_0^2} \Delta n_{gb}^2 \text{ and } \gamma_p = \frac{p}{\frac{4}{3}\pi r_p^3} \frac{3r_p\pi^2}{\lambda_0^2} \Delta n_p^2 \tag{5}$$

where r_{gb} (r_p) is the radius of the grain (pore); Δn_{gb}^2 (Δn_p^2) is the difference in the refractive indices between grains in birefringent materials (pore and matrix); p is the total porosity of the material and λ_0 is the wavelength of light used in vacuum. More recently, this analysis was extended to a distribution of pore and particle sizes (Suarez et al., 2009). In Figure 4.5, the effect of considering a distribution of pores instead of a single pore is given. For this simulation, a Poisson pore size distribution with an average radius of 30 nm was taken into account. It can be observed that the main differences take place at short wavelengths, that is, in the visible part of the spectrum for the numbers considered in this example.

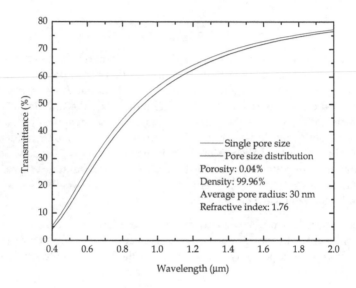

Fig. 4.5. Transmittance spectra in a material with a single pore size or a Poisson distribution of pore sizes.

In Figure 4.6, the effect of changing both the porosity for a fixed pore size (4.6.a) and the pore size at a fixed porosity (4.6.b) is simulated according to the formalism developed by Suarez et al. (Suarez et al., 2009). Some remarkable features should be highlighted: as shown in Figure 4.6.a, for a pore radius of 10 nm, in a 1 mm thick material and at a wavelength of 600 nm, i. e., in the visible range, a porosity of 0.5%, which corresponds to a density of the material of 99.5%, implies that the RIT decays to almost zero. Actually, values of the porosity over 0.3% lead to RIT values so small that their detection becomes very complicated for usual detectors. On the other hand, at the same wavelength and considering a porosity of only 0.05%, which corresponds to a density of 99.95%, it is shown that pore sizes over, approximately, 50 nm lead to very small RIT values. These two results combined imply that achieving a high density (>99.9%) is not usually enough to obtain a transparent material. It is also necessary to keep the values of pores below a given value that, according to these simulations, can be established below 30 nm.

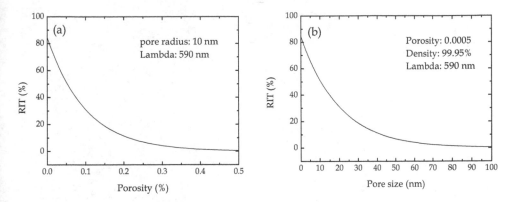

Fig. 4.6. Simulations considering different porosities and pore sizes.

The simulations shown above indicate the critical role of pores when transparent ceramics are being pursued. However, the situation can become more complex when the crystalline structure of the material considered is not cubic. In this case, the diagonal elements of the dielectric tensor will be, in general, different. In other words, the refractive index will depend on the crystalline orientation of each individual grain. For example, due to symmetry reasons, in alumina, two of the diagonal elements of the dielectric tensor are equal, but the third one is different ($\varepsilon_{xx}=\varepsilon_{yy}\neq\varepsilon_{zz}$). In alumina, this leads to a difference in the refractive index of only 0.008, but enough to induce a large scattering, particularly in the visible range. As shown in Figure 4.7a, the larger the grain size, the larger the scattering and, therefore, the smaller the RIT. Also, analogously to the porosity, the scattering becomes more critical at shorter wavelengths. It is shown in Figure 4.7a that whereas the RIT at 400 nm for a polycrystalline material does not reach 10%, it is close to 40% at 700 nm. Therefore, the grain size will be very important when materials are developed for the visible range and will not be so critical in devices operating in the infrared range. Following the same formalism, it can be shown that the RIT only falls to 60% at 1.5 microns and 70% at 2 microns when the average grain size is 2 microns in polycrystalline alumina. All RIT values shown so far have been calculated considering texture less materials. However, different routes have shown the possibility to induce some texture in the polycrystalline materials (Salem et al., 1989; Mao et al., 2008; Uchikoshi et al., 2004). Figure 4.7b shows how at two wavelengths, 400 and 700 nm, the RIT increases with increasing texture. In this simulation, texture = 0 corresponds to a perfectly polycrystalline material whereas texture = 1 corresponds to a fully textured, analogous to a single crystal, material. It is worth noting that no porosity has been considered in the simulations and that even so, at 400 nm, the RIT strongly decays from its maximum, around 80%, to only 20% due to a random orientation of the crystallites forming the material. The behaviour at 700 nm is similar, but the decay is not so abrupt, indicating again the strong wavelength dependence of the phenomenon. For these simulations, a grain size of only 500 nm has been considered, which is not easy to obtain in conventionally sintered alumina materials. Usual grain sizes are of the order of several µm, which leads to even more drastic reductions in the RIT, as can be deduced from Figure 4.7a.

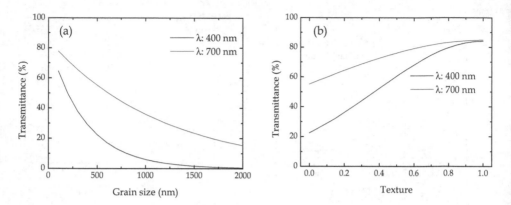

Fig. 4.7. Simulations for birefringent materials with different textures and grain sizes.

C. Pecharroman et al. developed a theoretical model (Pecharroman et al., 2009) for light scattering due to polycrystalline aggregates of uniaxial spheres within the Rayleigh-Gans-Debye approximation. This model shows that the scattering efficiency of each individual grain depends linearly on the grain size and on the texture of the samples. As indicated above, the most critical scattering is that due to the pores and this explains why common technical alumina ceramics which are considered dense at relative density over 99.5 % are opaque (white) when sintered in air, even if high purity raw materials are used. However, provided that the scattering due to pores decreases as $(d/\lambda)^4$, in order to make a material transparent it is not necessary to completely remove all the scattering centres, which is a cumbersome task, but it is enough to keep them all below a certain critical size. As an approximate rule, it is considered that the defects should be smaller than $\lambda/10$. In particular, for the visible range, in which the considered wavelengths range from 400 to 750 nm, the pore size should not be over a few tens of nm. Even so, the total porosity should never be over 0.05%. For this reason care must be taken not only during sintering, but also during processing and preparation of the green bodies. A uniform packing of the powders in green bodies has been shown in the literature to be critical to obtain low porosity ceramics. This is of particular importance when working with nanoceramics, as the attractive Van der Waals forces between particles are responsible for the high tendency of the nanoparticles to agglomerate. The presence of agglomerates leads to local differences in density which changes the sintering behaviour of the body leading to microcracks and microstructural effects such as residual pores and local coarsening and, therefore, to a large scattering. On the one hand, the raw powder should be very fine-grained in order to increase the sintering activity resulting in the elimination of residual porosity. Also, the structural homogeneity of sintering bodies is most important for a minimum of flaws that are detrimental to both optical and mechanical properties. This request for homogeneity makes the use of nanopowders < 100 nm still difficult. Casting methods are particularly suited to process raw powders with particle sizes between 100 and 150 nm.

Finally, it is obvious that preparing materials with the same thickness for a set of measurements is extremely difficult. For this reason, it is important to normalize measurements to a given thickness. The expression that relates the transmittance at a normalized thickness d_1 considering a measurement performed on a material with a thickness d_2 is given by

$$\frac{T(d_1)}{(1-R_s)^2} = e^{-\gamma d_1} = e^{-\gamma d_1 \cdot \frac{d_2}{d_2}} = \left[e^{-\gamma d_2} \right]^{d_1/d_2} = \left[\frac{T(d_2)}{(1-R_s)^2} \right]^{d_1/d_2} \quad (6)$$

5. Conclusions

In conclusion, the interest on transparent ceramic materials relies on the extraordinary combination of high mechanical performance, chemical resistance/inertness... and little or none absorption in different ranges of the electromagnetic spectrum from infrared to near ultra violet. It has been shown that obtaining a transparent ceramic material implies sintering to a density, very close to the theoretical density ($\geq$99.9%), and keeping the few pores left with a pore size below 30 nm. In order to achieve this aim, attention must be paid not only to the raw materials: adequate granulometry, particle size... but it becomes also critical to control the packing of the powders in the green body. Usually, large defects present in the green bodies cannot be removed during the sintering process, and therefore, forming step is decisive for the preparation of transparent materials. Moreover, it has been shown how an adequate choice of the sintering technique combined with a tailoring of the starting powder is critical when an accurate control of the microstructural features is required. Finally, different analysis techniques of the transmittance spectra have been given from which average structural parameters such as pore size, porosity and grain size and texture in birefringent ceramics can be extracted.

6. References

Álvarez-Clemares I., Mata-Osoro G., Fernández A., López-Esteban S., Pecharromán C., Palomares J., Torrecillas R. & Moya J.S. (2010). Transparent alumina/ceria nanocomposites by spark plasma sintering. Advanced Engineering Materials, 12, 11, 1154-1160

Amato I., Baudrocco F. & Martorana D. (1976). Evaluation of freeze drying and spray drying processes for preparing transparent alumina. Material Science and Engineering, 26, 1, 73-78.

Anderson R.C. (1998). Transparent Yttria-based ceramics and method for producing same. U. S. Patent 3545987.

Appetz R. & Van Bruggen M.P.B. (2003). Transparent alumina: a light scattering model. Journal of the American Ceramic Society, 86, 3, 480-486.

Bennison S.J. & Harmer M.P. (1990a). A history of the role of MgO in the sintering of alfa-alumina. Ceramic Transactions, American Ceramic Society, 7, 13-49.

Bennison S.J. & Harmer M.P. (1990b). Effect of magnesia on surface diffusion in sapphire and the role of magnesia in the sintering of alumina. Journal of the American Ceramic Society, 73, 4, 833-837

Bohren C. F. & Huffmann D.R. (2010). Absorption and scattering of light by small particles, Wiley-Interscience, ISBN 3527406646, New York

Born M. & Wolf E. (2005). Principles of Optics-Electromagnetic theory of propagation, interference and diffraction of light, Cambridge University Press, ISBM 0521642221, Cambridge, UK.

Budworth D.W. (1970). The selection of grain-growth control additives for the sintering of ceramics. Mineralogical Magazine, 37, 291, 833-838.

Chaim R., Marder-Jaeckel R. & Shen J.Z. (2006). Transparent YAG ceramics by surface softening of nanoparticles in spark plasma sintering. Material Science and Engineering, 429, 1-2, 74-78

Cheng J., Agrawal D., Zhang Y. & Roy R. (2002). Microwave sintering of transparent alumina. Materials Letters, 56, 4, 587– 592.

Chen Z., Li J., Xu J. & Hu Z. (2008). Fabrication of YAG transparent ceramics by two-step sintering. Ceramics International, 34, 7, 1709-1712

Coble R. L. (1962). U.S. Pat. No.3 026 210.

Cook R., Kochis M., Reimanis I., Kleebe H. J. (2005). A new powder production route for transparent spinel windows properties: powder synthesis and windows properties. Proceedings of the Defense and Security Symposium.

Dalisa, A. L. & Seymour R. J. (1973). Convolution Scattering Model for Ferroelectric Ceramics and Other Display Media, Proceedings of the IEEE, 61, 7, 981–991.

Greskovich C. & Chernoch J.P. (1973). Polycrystalline ceramic lasers. Journal of Applied Physics, 44, 4599-4607.

Handwerker C.A., Morris P. A. & Coble R. L. (1989). Effects of chemical inhomogeneities on grain growth and microstructure in Al_2O_3. Journal of The American Ceramic Society, 72, 1, 130-136.

Hartnett T. M., Bernstein S. D., Maguire E. A., & Tustison R. W. (1998). Optical properties of AlON (aluminum oxynitride). Infrared Physics & Technology, 39, 4, 203–211.

Ikesue A., Furusato I. (1995). Fabrication of polycrystalline transparent YAG ceramics by a solid-state reaction method. Journal of the American Ceramic Society, 78, 1 225-228.

Itatami K., Tsujimoto T. & Kishimoto A. (2006). Thermal and optical properties of transparent magnesium oxide ceramics fabricated by post hot-isostatic pressing. Journal of the European Ceramic Society, 26, 4-5, 639-645

Kim B. N., Hiraga K., Morita K. & Yoshida H. (2007). Spark Plasma Sintering of transparent alumina. Scripta Materialia, 57, 7, 607-610

Kong, J., Tang, D.Y., Zhao B., Lu J., Ueda K., Yagi H. & Yanagitani T. (2005). 9.2-W diode-pumped $Yb:Y_2O_3$ ceramic laser. Applied Physics Letters, 86, 16, 16116-16119.

Mccauley J. W., Patel P., Chen M., Gilde G., Strassburger E., Paliwal B., Ramesh K.T. & Dandekar D.P (2099). AlON: A brief history of its emergence and evolution, Journal of the European Ceramic Society, 29, 2, 223-236.

Li J. G., Ikegami T., Lee J. H. & Mori T. (2000a). Low-temperature fabrication of transparent yttrium aluminum garnet (YAG) ceramics without additives. Journal of the American Ceramic Society, 83, 4, 961-963.

Li J. G., Ikegami T., Lee J. H., Mori T. & Yajima Y. (2000b). Co-precipitation synthesis and sintering of yttrium aluminum garnet (YAG) powders: the effect of precipitant. Journal of The European Ceramic Society, 20, 14-15, 2395-2405.

Lu T. C., Chang X. H., Qi J. Q., Lu X. J., Wei Q. M., Zhu S., Sun K., Lian J. & Wanga L. M. (2006). Low-temperature high-pressure preparation of transparent nanocrystalline $MgAl_2O_4$ ceramics. Applied Physics Letters, 88, 21, 213120-213123.

Lukin E. S., Vlasov A. S., Zubakhina M.A. & Datsenko A.M. (1980). Effect of γ radiation on the optical properties of transparent yttrium-oxide ceramics. Glass and Ceramics, 37, 5, 255-258.

Mah T. I., Parthasarathy T. A. & Lee H. D. (2004). Polycristalline YAG; structural of functional?. Journal of Ceramic Processing Research, 5, 4, 369-379.

Mao X., Wang S., Shimai S. & Guo J. (2008). Transparent Polycrystalline Alumina Ceramics with Orientated Optical Axes, Journal of the American Ceramic Society. 94, 10, 3431-3433.

Mie G. (1908). Beiträge zur Optik trüber Medien, speziell kolloidaler Metallösungen, Annalen der Physik. 330, 377–445.

Morita K., Kim B.N., Hiraga K., Yoshida H. (2008). Fabrication of transparent $MgAl_2O_4$ spinel polycrystal by spark plasma sintering processing. Scripta Materialia, 58, 12, 1114–1117.

Mouzon, J., Glowacki E. & Odén M. (2008). Comparison between slip-casting and uniaxial pressing for the fabrication of translucent yttria ceramics. Journal of Material Science, 43, 8, 2849-2856

Patel P.J., Gary A. G., Dehmer P. G. & McCauley J.W. Transparent Armor, The AMPTIAC Newsletter. Advanced Materials and Processes Technology, 4, 3, 1-24.

Pecharroman C., Mata-Osoro G., Diaz L. A., Torrecillas R. & Moya J. S. (2009). On the transparency of nanostructured alumina: Rayleigh-Gans model for anisotropic spheres, Optics Express 17, 8, 6899-6912.

Peelen J. G. J. & Metselaar R. (1974). Light Scattering by Pores in Polycrystalline Materials: Transmission Properties of Alumina, Journal of Applied Physics, 45, 1, 216-220.

Peuchert U., Okano, Y., Menke Y., Reichel S. & Ikesue, A. (2009). Transparent cubic-ZrO_2 ceramics for application as optical lenses. Journal of the European Ceramic Society, 29, 2, 283-291.

Qin G., Lu J., Bisson J. F., Feng Y., Ueda K. I., Yagi H. & Yanagitani T. (2004). Upconverion luminescence of Er^{3+} in highly transparent YAG ceramics. Solid State Communications, 132, 2, 103-106.

Salem J. A., Shannon Jr J. L. & Brad R. C. (1989). Crack Growth Resistance of Textured Alumina, Journal of the American Ceramic Society. 72, 1, 20-27.

Savastru D., Miclos S., Cotïrlan C., Ristici E., Mustata M., Mogïldea M., Mogïldea G., Dragu T. & Morarescu R. (2004). Nd:YAG Laser system for ophthalmology: Biolaser-1. Journal of Optoelectronics and Advanced Materials, 6, 2, 497–502.

Shikao S. & Jiye W. (2001). Combustion synthesis of Eu activated $Y_3Al_5O_{12}$ phosphor nanoparticles. Journal of Alloys and Compounds, 327, 1-2, 82–86.

Shuzhi L., Bangwei Z., Xiaolin S., Yifang O., Haowen X. & Zhongyu X. (1999). The structure and infrared spectra of nanostructured $MgO-Al_2O_3$ solid solution powders prepared by the chemical method. Journal of Materials Processing Technology, 89-90, 405-409.

Suarez M., Fernandez A., Menendez J.L. & Torrecillas R. (2009). Grain growth control and transparency in spark plasma sintered self-doped alumina materials. Scripta Materialia 61, 10, 931-934.

Suárez M., Fernández A., Menéndez J.L., Nygren M., Torrecillas R. & Zhao Z. (2010). Hot Isostatic pressing of optically active Nd:YAG powders doped by a colloidal processing route. Journal of the European Ceramic Society, 30, 6, 1489-1494

Suárez M., Fernández A., Menéndez J.L. & Torrecillas R. (2011). Blocking of grain reorientation in self-doped alumina materials. Scripta Materialia, 64, 6, 517-520

Takaichi K., Yahi H., Lu J., Shirakawa A., Ueda K., Yanagitani T. & Kaminskii A.A. (2003). Yb^{3+} doped $Y_3Al_5O_{12}$ ceramics- A new solid-state laser material. Physics state solid A, 200, 1, R5-R7.

Tokurakawa, M.T., Kazunori S., Akira U., Ken-ichi Y., Hideki Y., Takagimi K. & Alesander A. (2007). Diode-pumped 188 fs mode-locked Yb^{3+}:Y_2O_3 ceramic laser, Applied Physics Letters, 90, 7, 071101-071104.

Uchikoshi T., Suzuki T.S., Okuyama H. & Sakka Y. (2004). Control of crystalline texture in polycrystalline alumina ceramics by electrophoretic deposition in a strong magnetic field, Journal of Materials Research, 19, 5, 1487-1491.

Van de Hulst H. C. (1957). Light scattering by small particles, John Wiley and Sons, ISBN 0486642283, New York.

Wang C. & Zhao Z. (2009). Transparent $MgAl_2O_4$ ceramic produced by spark plasma sintering. Scripta Materialia, 61, 2, 193-196

Wei C. G. (2009). Transparent ceramis for lighting. Journal of the European Ceramic Society, 29, 2, 237-244.

Wen L., Sun X., Xiu Z., Chen S. & Tsai C .T. (2004). Synthesis of nanocrystalline yttria powder and fabrication of transparent YAG ceramics. Journal of the European Ceramic Society, 24, 2681-2688.

Zhang G., Wang Y., Fu Z., Wang H., Wang W., Zhang J., Lee S.W. & Niihara K. (2009). Transparent mullite ceramic from single-phase gel by spark plasma sintering. Journal of the European Ceramic Society, 29, 13, 2705-2711

6

Reactive Sintering of Aluminum Titanate

Irene Barrios de Arenas
Instituto Universitario de Tecnología
"Dr Federico Rivero Palacio"
Venezuela

1. Introduction

The high thermal shock resistance due to the negligible thermal expansion coefficient, additional to its low thermal conductivity and good chemical resistance makes the aluminum titanate (Al_2TiO_5) a suitable material for different technological applications. It is a ceramic material consisting of a mixture of alumina (Al_2O_3) and titania (TiO_2) forming solid solution with stoichiometric proportion of the components: $Al_2O_3 \cdot TiO_2$ or Al_2TiO_5. It is prepared by heating a mixture of alumina and titania at temperature above 1350°C, in air atmosphere. Pure Aluminum Titanate is unstable at temperatures above 750°C, when the solid solution decomposes, following a eutectoid reaction, into two separate phases Al_2O_3 and TiO_2. For this reason Aluminum Titanate ceramics are doped usually with MgO, SiO_2 and ZrO_2 in order to stabilize the solid solution structure.

Unfortunately, the expansion crystal structure anisotropy that promotes the low thermal expansion coefficients provokes microcracking, as a result of anisotropy of thermal expansion along the three primary axes of the crystal lattice (a single crystal of Aluminum Titanate expands along two axes and contracts on the third one when heated), therefore a low mechanical strength and, on the other hand, the low thermal stability below 1280°C restricts its technical use.

Aluminum Titanate ceramic materials have many technological applications, among others, as thermal insulation liner, soot particulate filter in diesel engines, spacing rings of catalytic converters, in the foundry crucibles, launders, nozzles, riser tubes, pouring spouts and thermocouples for non-ferrous metallurgy and master moulds glass industries.

2. Fundamentals of low thermal expansion coefficient aluminum titanate (Al_2TiO_5) ceramics

There are two important features to achieve a very low thermal expansion coefficient, in crystalline ceramic structures highly anisotropic. The first aspect involves unit cell crystalline chemistry. The coefficients of thermal expansion of the crystal axes are controlled to develop solid solutions, in an attempt to ensure that the sum of the coefficients of the principal axes gives zero. In the case of polycrystalline ceramic materials, the volumetric thermal expansion coefficient is related to the sum of unit cells coefficients of thermal expansion. In orthorhombic crystal structures as that of the pseudobrookita, material object of this work, the relationship is:

$$\beta_v = \alpha_a + \alpha_b + \alpha_c \tag{1}$$

Where β_v = Volumetric thermal expansion coefficient
          α_i = Thermal expansion coefficients of principal crystal axes

As the anisotropic crystalline structures have principal axes with positive and negative expansion coefficients, it is necessary to examine the thermal expansion coefficients of all the members of an isostructural family and chemically design a solid solution whose α_i addition is close to zero. Bayer (1971; 1973), studied the unit cell, of the pseudobrookita structure. Provided that the sum of the thermal expansion coefficients of the principal axes (α_i) add zero, it occurs an inevitable combination of positive and negative values. This condition leads to, very high (at GPa levels), micromechanical stresses at grain boundaries, during cooling from the temperatures of ceramic processing. The development of these internal stresses, promotes the breakdown of the grain boundaries, which causes a decrease in the structural integrity of the polycrystalline ceramic body. However, the existence of this microcracking depends on the microstructural grain size. Kuszyk and Bradt (1973) noted that the rigidity of the ceramic body decreased as increasing grain size, determining a critical grain size. Once determined this size, is simply necessary a process production control to achieve a compromise between the microcracking and the required structural mechanical resistance. Another possibility is to produce a material with large grain size and extensive microcracking with low mechanical resistance but where the main interest is the low thermal expansion (Hasselman, 1977; Stingl, 1986; Sheppard 1988; Huber, 1988). However, several researchers (Buessem, 1966; Cleveland, 1977; 1978) have suggested that the presence of the extensive internal microcracking, contributes to an increase in the resistance to fracture of these polycrystalline ceramics highly anisotropic, activating mechanisms such as: shielding, branching or cracks deviation. Experimentally, this hypothesis has not been demonstrated, so it is a concept that must be handled carefully.

3. Aluminum titanate (Al$_2$TiO$_5$) ceramics

The Al$_2$TiO$_5$ is a one mole Al$_2$O$_3$ and one mole TiO$_2$ compound. This material is conventionally prepared by reactive sintering of Al$_2$O$_3$ and TiO$_2$ powders. Its interest as polycrystalline ceramic material arose from the work of Bachmann (1948), who found that the thermal expansion of aluminum titanate, in the studied temperature range, could be lower than that of the vitreous silica. This material has interesting features for applications such as thermal insulator and can also withstand strong thermal gradients. This aluminum titanate emerged as a promising ceramics for technological applications; summarizing its most important physical properties, in Table 1.

The material presents two major problems: the thermodynamics instability of the Al$_2$TiO$_5$ below 1280 °C and its poor mechanical resistance related to an extensive microcracking which is, in turn, responsible for the low thermal expansion. Decomposition can be controlled or at least delayed, with oxides additions, such as MgO (Ishitsuka and col., 1987;

Wohlfrom and col., 1990) and Fe_2O_3 (Tilloca, G., 1991, Brown et al., 1994), which forms solid solutions between the Al_2TiO_5 and the isoestructurals $MgTi_2O_5$ and Fe_2TiO_5. The mechanical strength can be increased with good results preparing composite materials such as: Al_2TiO_5 - Mulita (Morishima and col. 1987), Al_2TiO_5 - Mulita - ZrO_2 (Wohlfrom et al., 1990).

Property	Al_2TiO_5	Reference
Density (g/cm^3)	3.702	Holcombe (1973)
Thermal Expansion Coefficient Average $(x10^{-6}\ ^{\circ}C^{-1})$ $\alpha_{a20-520} - \alpha_{a20-1000}$ $\alpha_{b20-520} - \alpha_{b20-1000}$ $\alpha_{c20-520} - \alpha_{c20-1000}$	-2.9 - -3 10.3 - 11.8 20.1 - 21.8	Wohlfromm (1990)
Thermal Expansion Coefficient Average $(x10^{-6}\ ^{\circ}C^{-1})$ Crystallographic $\alpha_{20-520} - \alpha_{20-1000}$ Macroscopic $\alpha_{20-1000}$ $\alpha_{20-1000}$ Anisotropy $\Delta\alpha_{20-520} - \Delta\alpha_{20-1000}$	 9.2 - 10.2 1.0 - 1.5 1.5 - 1.7 23 -24.8	Stingl (1986) Milosevski (1995)
Melting Point $(^{\circ}C)$	1860	Lang (1952)
Elastic Modulus E(GPa)	12 - 18 10 - 20 13 - 15	Stingl (1986) Cleveland (1978) Milosevski (1997)
Hardness, Hv (GPa)	5	Wohlfromm (1990)
Bending Strenght, σ (MPa)	4 - 20 25 - 40	Milosevski (1995)
Thermal shock resistance (Wm^{-1})	500	Stingl (1986)
Thermal Conductivity, k(W/mK)	1.5 -2.5	Stingl (1986) Milosevski (1997)

Table 1. Aluminum Titanate Physical Properties.

3.1 Equilibrium diagram

Lang et al. (1952) studied the Al_2O_3 -TiO_2 equilibrium diagram (Fig. 1), finding the existence of two allotropic forms of aluminum titanate: α- Al_2TiO_5, a high temperature phase, stable between 1820°C and the melting point at $1860\pm10°C$ and β-Al_2TiO_5, a low temperature phase stable from room temperature up to $\approx 750°C$ and from 1300°C up to inversion temperature 1820°C (at intermediate values, it has instability and decomposes to Al_2O_3 + TiO_2). The transformation between both phases is spontaneous and reversible; it was found that it is almost impossible to obtain α-Al₂TiO₅ at room temperature, being necessary cooling speeds greater than 800 K/h.)

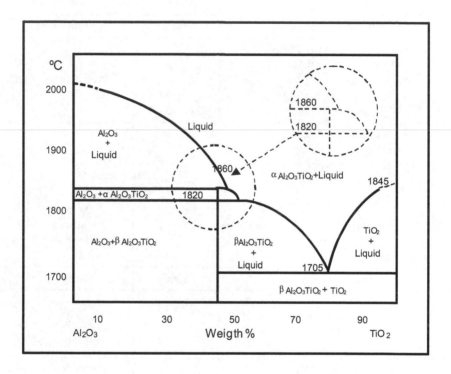

Fig. 1. Al₂O₃ -TiO₂ Equilibrium Diagram by Lang (1952).

Evidence suggests a congruent transformation of α- Al₂TiO₅ at 1860°C, but the possibility of an incongruent fusion or the existence of a solid solution between Al₂O₃ and Al₂TiO₅ could not be studied, due to the difficulties of obtaining accurate data, because of the high viscosity of liquid formed.

A second point of importance obtained from this research was the suggestion of an instability region for the β- aluminum titanate between 750°C and 130°C. This phenomenon has been confirmed by subsequent research (Fonseca & Baptista 2003). Lang et al., (1952) concluded that opposing formation and decomposition processes occurred in dynamic equilibrium, promoting decomposition at a certain temperature range, i.e. the β- Al₂TiO₅ phase is stable above 1300°C, below this temperature aluminum titanate undergoes an eutectoid transformation according to the reaction:

$$Al_2TiO_5 \Leftrightarrow Al_2O_3 + TiO_2 \tag{2}$$

3.1.1 Effect of oxygen partial pressure in aluminum titanate stability

The composition and mechanical properties of aluminum titanate are strongly influenced by the partial pressure of oxygen from the surrounding atmosphere. It is well known that the valence of the Ti cation in titanium oxides depends on the partial pressure of oxygen (Jürgen et al. 1996). At high oxygen pressures, Ti is tetravalent producing TiO_2. Due to entropic reasons, at low oxygen pressures (such as for example in air) there is a small fraction of Ti^{+3}, its amount depending on the temperature. By decreasing the partial pressure of oxygen furthermore, the Ti^{+3}/Ti^{+4} relationship increases continuously producing Ti_nO_{2n-1} Magneli phases and other sub-oxides such as Ti_4O_7, Ti_3O_5 and Ti_2O_3. The aluminum titanate phase shows similar behavior (Jürgen et al. 1996). Under high O_2 pressure the aluminum titanate is almost a stoichiometric composition phase Al_2TiO_5. Decreasing the oxygen partial pressure, due to the increase in the Ti^{+3}/Ti^{+4} ratios, it occurs a gradual interchange of Al^{+3} by Ti^{+3} in the aluminum titanate structure. This exchange results in the stoichiometric Al_2TiO_5 decomposition, to the reduced form of aluminum titanate, Al_2O_3 and oxygen according to the reaction:

$$(3-2z)[(Al^{+3})_2(Ti^{+4})_1(O^{-2})_5] = (Al_z^{+3}Ti_{1-z})_2(Ti^{+4})(O^{-2})_5 + 3(1-z)Al_2O_3 + \tfrac{1}{2}(1-z)O_2 \qquad (3)$$

Decreasing the oxygen potential ($z \to 0$), the decomposition reaction eventually produces Ti_3O_5 titanium oxide. The degree of decomposition from Al_2TiO_5 to Ti_3O_5 can be related to a continuous change of the aluminum titanate parameter network c, (Asbrink et al., 1967).

Considering the solubility between the Al_2TiO_5 and the Ti_3O_5 under various oxygen pressures, the Al_2TiO_5 was described with a subnet model $(Al^{+3}, Ti^{+3})_2(Ti^{+4})_1(O^{-2})_5$, this model was derived from the Al_2TiO_5 orthorhombic structure (Epicer et al. 1991), taking into account the mutual exchange of trivalent cations in one subnet, while in the other subnet occupied by Ti^{+4} and O^{-2} species there is no influence.

Subsequently, Freudenberg (1987), brings together all the data obtained and proposed a modified diagram (Fig. 2). Where the only stable compound in the Al_2O_3 – TiO_2 system is considered to be, the β- Al_2TiO_5 phase; this compound decomposes above 1280 ± 1°C (Kato et al. 1980).

The Al_2TiO_5 divides the system in two sub-systems Al_2O_3 - Al_2TiO_5 and Al_2TiO_5 - TiO_2 with eutectics at titania 38.5 and 80 weight percent respectively (Fig. 2).

It is important to point out the remarkable solubility difference between corundum and titania, the Al_2O_3 in TiO_2 is ≈ one order in magnitude higher than the TiO_2 in Al_2O_3, and the Al_2O_3 and TiO_2 solubility are practically null in Al_2TiO_5; so it is considered aluminum titanate as a stoichiometric compound. However this claim is only correct for oxidizing atmospheres.

The Al_2O_3 has a 2.5% molar maximum solubility in TiO_2 (1.97 ± 0.18 in weight) at 1726°C, (Slepetys, 1969), whereas the solubility of the latter in the alumina is almost non-existent 0.35% molar between 1300 and 1700°C. While both oxides solubility in the Al_2TiO_5 is completely null (Golberg, 1968).

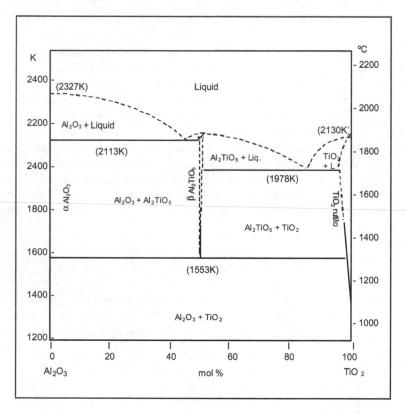

Fig. 2. Al_2O_3 -TiO_2 Equilibrium Diagram calculated in air, from experimental review by Freudenberg (1987).

4. Additives

The thermal instability of aluminum titanate and its low mechanical resistance are the main reasons for the additives use, taking into account these will influence the production process and the final product properties. An important characteristic for all additives is that they do not decrease significantly aluminum titanate thermomechanical properties. Small additions ($\leq$ 5% by weight) are usually added with the aim of forming aluminum titanate solid solutions.

As was mentioned before, the aluminum titanate is formed above and decomposes below the equilibrium temperature 1280°C (Kato et al., 1980), with a free energy of formation given by:

$$\Delta G^\circ \ Al_2TiO_5 = \Delta H^\circ - \Delta S^\circ T \qquad (4)$$

$$\Delta G^\circ \ Al_2TiO_5 = 17000 - 10.95T \qquad (5)$$

The endothermic reaction is possible due to the entropy ($\Delta S°$) positive contribution. So as other pseudobrookitas, Al_2TiO_5 can be stabilized entropically (Navrotsky 1975), with certain contributions to cation disorder (Morosin et al., 1972). It is conceivable that the positive effect of entropy can be reinforced with additional entropy in terms of mixing by the formation of aluminum titanate solid solutions. It has been determined empirically that solid solutions containing Fe^{+3} and Mg^{+2}, provide a lower decomposition temperature, i.e. increasing stability. On the other hand, solid solutions with Cr^{+3} promote a greater temperature of decomposition, i.e., reducing stability (Woermann 1985).

Jung et al. (1993), studied the replacement of Ti^{+4} by Ge^{+4} and Al^{+3} by Ga^{+3} and Ge solid solutions combined also with additions of MgO and Fe_2O_3, finding that the stabilizing effect of the additions decreased in the following order: Fe^{+3}, $Mg^{+2} > Ge^{+2} > Ga^{+3}$, corroborating data found in previous research that Fe^{+3}, Mg^{+2} are the best stabilizers so far.

Additions such as Fe_2O_3, MgO or SiO_2 were studied, the first two promoting structures of the pseudobrookites type Fe_2TiO_5 and $MgTi_2O_5$ giving complete solid solutions with Al_2TiO_5 (Brown 1994; Buscaglia et al., 1994; 1995; 1997). The SiO_2 has limited solubility (Ishitsuka 1987), however additions up to 3 weight percent produce a slight increase in the mechanical resistance, due to small amounts of liquid phase that densify the material but, larger amounts cause excessive growth of the grain that is detrimental to the mechanical resistance, (Thomas et al., 1989).

Liu et al., (1996), studied the thermal stability of Al_2TiO_5 with Fe_2TiO_5 and $MgTi_2O_5$ additions finding that material with Fe^{+3} additions did not show any significant mechanical properties decomposition or degradation and the material with Mg^{+2} annealed to 1000 - 1100°C showed an Al_2O_3 and TiO_2 breakdown.

5. Experimental procedure

The raw materials used were reactive grade: Al_2O_3 (D_{50}=0.60μm), TiO_2 (D_{50}=0.88μm), V_2O_5 (D_{50}=0.60μm), MnO (D_{50}=0.60μm), ferrosilicon (D_{50}=0.69μm), $FeTiO_3$ (D_{50}=0.82μm), and, alumina ball milled 98.5%$FeTiO_3$-1.2%SiO_2 purified mineral (D_{50}=0.88μm).

Two (2) equimolar mixtures of Al_2O_3 and TiO_2 (56wt%Al_2O_3 - 44wt%TiO_2) were homogeneously mixed with 3, 6 and 9 wt% of each additive using alumina jars and balls, during 6 hours. No binder has been added to the aqueous media powder mixture and it was dried out at 120°C for 24 hours. The material was crushed in an alumina mortar prior to the manufacture of samples by uniaxial die compaction at 300 MPa. Green bodies were reactive sintered at 1450°C, in air for 3 h. Heating was programmed at 5°C/min. whereas cooling at 15°C/min, in order to avoid eutectoid transformation: $Al_2TiO_5 \rightarrow Al_2O_3 + TiO_2$ (Kolomietsev et al.,1981).

X-ray diffraction (XRD) analysis has been performed on powders from crushed sintered samples, with grains below 30 μm suitable to obtain rigid specimens. The quantification of Al_2TiO_5 formed was determined by the internal standard method, through direct determination based on the methodology of Klug and Alexander (1954). In this study, the diffraction signals used were: Al_2TiO_5 (023), Al_2O_3 (104) and TiO_2 Rutile (110), which are representative of the three components of interest in the studied samples.

Sintered sample surfaces were carefully ceramographically prepared to minimize damage and, in some cases it was needed to chemically etch in ambient 15%HF solution for 60 s, to reveal grain boundaries. The microstructure characterization was carried out using compositional back scattered electron images (BSEI) from scanning electron microscopy (SEM) and energy-dispersive X-ray spectroscopy (EDS). Evaluation of grain size and phases present has been performed by image analysis.

In order to quantify the stabilization of Al_2TiO_5, sintered samples previously thermal treated at 1100°C for 100 h. were Si internal standard XRD analyzed, as in the as-sintered condition.

To determine the type of Fe ion in solution, it was used Mössbauer Spectroscopy with the isotope iron ^{57}Fe, in the samples with addition of ilmenite and ferrosilicon. The source used was ^{57}Co, the Mösssbauer transition is 14.41 keV, with the excited level of nuclear spin I = 3/2 and fundamental level I = 1/2. The extent of the isomeric shift provides information on the valence of the atom to which belong the core, as the electronic layers and therefore the density of electrons in the nucleus, are sensitive to chemical bonding.

Thermal expansion analysis in the temperature range of 25 to 1000°C and 1450°C, at 5°C/min heating and cooling ramps, has been performed on selected samples with good stabilization behavior.

6. Discussions of results

6.1 Additives selection

The additives selection is based on the cation radius, which must be related to the aluminum titanate cations i.e., Al^{+3} and Ti^{+4}, in order to be able to replace them in the solid solution to be formed. The thermal expansion can be related to the degree of crystalline distortion which is known to increase with the difference between the radii of the cations.

Consequently, the octahedral distortion is much greater in Al_2TiO_5 than in Fe_2TiO_5, due to the small radius of Al^{+3} ions, which facilitate a tendency to the tetrahedral coordination (Bayer, G., 1973). To avoid extreme distortion, the replacing selected cations must have a radii at least close to the Al^{+3}=0.54 Å; i.e., V^{+5}=0.59 Å; Mn^{+4} =0.76 Å; Si^{+4} =0.41 Å; Fe^{+3}=0.67 and Ti^{+4}=0.76. Hence, all additives used in this work fulfill the requirement. On the other hand, the aluminum titanate is formed by an equimolar reaction between Al_2O_3 and TiO_2. However, due to entropic reasons, in low O_2 pressure conditions, as in air for example, there is a small fraction of Ti^{+3}, its quantity being a function of the temperature. Hence, the reaction of transformation can be expressed in terms of the intermediate reaction of the titanium oxide Ti_3O_5 in the following manner:

$$\alpha\ Al_2O_3 + 1/3\ Ti_3O_5 + 1/6\ O_2 \Leftrightarrow Al_2TiO_5 \tag{6}$$

The Ti_3O_5 can be seen in terms of $Ti^{+3} Ti^{+4} O_5^{-2}$ for which there is the possibility of forming "*limited solid solutions*" by cationic substitution as:

$$(1-x)Al_2TiO_5 + xTi_3O_5 \Leftrightarrow (Al^{+3}_{1-x}Ti^{+3}_x)_2Ti^{+4}O_5^{-2} \tag{7}$$

Studying the affinity diagram of multivalent oxides of the transition metals with O_2, it was observed that below Ti, the order of decreasing affinity with O_2 is V, Mn and Fe (Fig.3.).

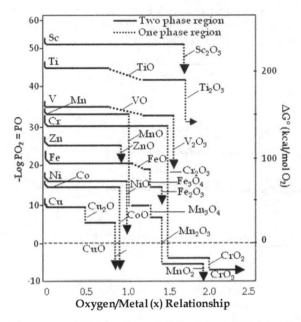

Fig. 3. Oxygen partial pressure (PO_2) vs. Oxygen/Metal (x en MO_x) for the 3d metals.

6.2 Structure analysis

The XRD results showed, for aluminum titanate without additive (Fig. 4.), that the selected temperature and time are sufficient for a near 100% Al_2TiO_5 reaction of formation, as the most important peaks correspond to this compound with a minimum of Al_2O_3 and TiO_2 remnants. It is important to point out that in all XRD, are represented the PDF values for all the constituents expected in each case, although they are not present.

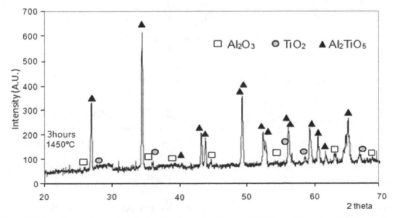

Fig. 4. X Ray Diffraction of equimolar mixture Al_2O_3 and TiO_2 without addition, sintered at 1450°C for 3 hours.

In the samples with V_2O_5, the formation of Al_2TiO_5 decreases as addition contents increase due to an intergranular liquid phase formed, identified by SEM-EDX, that inhibits the reaction between the main constituents (Fig. 5.)

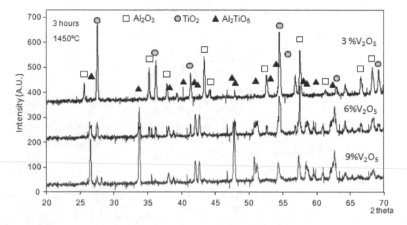

Fig. 5. X Ray Diffraction of equimolar mixture Al_2O_3 and TiO_2 with V_2O_5: 3, 6 and 9 wt% addition, sintered at 1450°C for 3 hours.

Regarding the MnO addition, it promotes Al_2TiO_5 formation with contents, as depicted by the aluminum titanate principal signals which increase in intensity whereas those of Al_2O_3 and TiO_2 are suppressed. $MnTiO_3$ appears as product of TiO_2 and MnO eutectic reaction. (Fig. 6.).

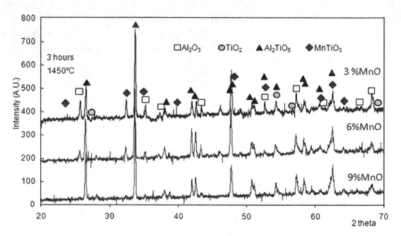

Fig. 6. X Ray Diffraction of equimolar mixture Al_2O_3 and TiO_2 with MnO: 3, 6 y 9 wt% addition, sintered at 1450°C for 3 hours.

In the case of ferrosilicon added compositions (Fig. 7), the Al_2TiO_5 formation reaction occurs but not complete and, the main signals of Al_2O_3 and TiO_2 are slightly shifted, corresponding to Al_2SiO_5 and $Al_4Ti_2SiO_{12}$ being the latter, a product of a ternary eutectic transformation.

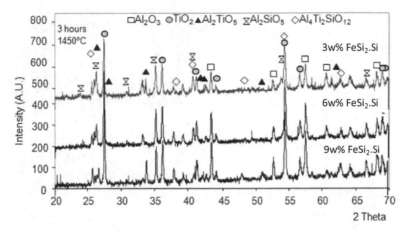

Fig. 7. X Ray Diffraction of equimolar mixture Al_2O_3 and TiO_2 with industrial $FeSi_2.Si$: 3, 6 and 9 wt% addition, sintered at 1450°C for 3 hours.

Both pure and concentrated mineral ilmenite ($FeTiO_3$) (Figs.8 and 9),promoted the formation of Al_2TiO_5 in all compositions studied. Phases such as Fe_2O_3, TiO_2, or Fe_2TiO_5 product of decomposition and reaction of the $FeTiO_3$, due to the oxidizing atmosphere, were not detected.

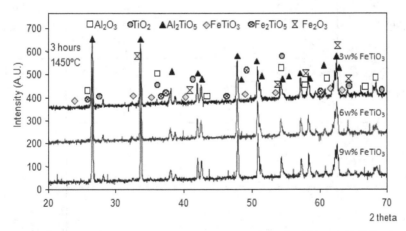

Fig. 8. X Ray Diffraction of equimolar mixture Al_2O_3 and TiO_2 with pure $FeTiO_3$: 3, 6 y 9 wt% addition, sintered at 1450°C for 3 hours.

Other remark is that expected phases, product of the reaction of contaminant SiO_2 with the parent Al_2O_3 and TiO_2, in the samples containing mineral did not show in the XRD spectra. Nevertheless, the most important reflections correspond to the Al_2TiO_5 corroborating the beneficial effect of this additive in its formation (Fig. 9).

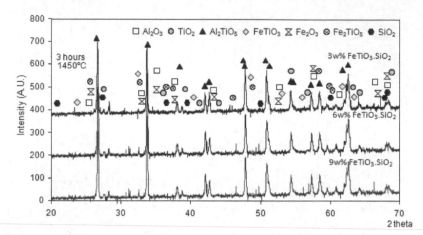

Fig. 9. X Ray Diffraction of equimolar mixture Al_2O_3 and TiO_2 with concentrated placer ilmenite ($FeTiO_3.SiO_2$): 3, 6 and 9 wt% addition, sintered at 1450°C for 3 hours.

6.2.1 Al_2TiO_5 formation phase quantification

The quantification of Al_2TiO_5 formed, was determined by the internal standard method; the achieved results are showed in Table 2.

AT+Additive (%)	%	%TiO_2 unreacted	% Al_2O_3 unreacted	%Al_2TiO_5 formed
	3	19.2	24.3	56.5
V_2O_5	6	20.1	25.5	54.4
	9	22.6	28.7	48.7
	3	13.1	16.6	70.4
MnO	6	11.2	14.1	74.7
	9	9.4	11.9	78.7
	3	5.5	5.8	88,7
$FeTiO_3 .SiO_2$	6	5.1	6.4	88.5
(mineral)	9	5.3	628	88.5
	3	1.9	2.7	95,0
$FeTiO_3$	6	2.3	2.8	94,8
(pure)	9	2.1	2.0	96,0
	3	21.3	27.1	51.7
$FeSi_2$	6	19.8	25.1	55.1
	9	16.9	21.5	61.6

Table 2. Al_2TiO_5 % phase formation by sintering at 1450°C/3hours.

It can be seen that the best results are obtained firstly for the pure ilmenite, secondly the mineral ilmenite, then the MnO, vanadium oxide and ferrosilicon additions respectively.

6.3 Microstructure analysis

The composition without addition (Fig. 10), shows the characteristic microstructure of the aluminum titanate: a porous and microcracked Al_2TiO_5 matrix phase and the presence of unreacted Al_2O_3 and TiO_2, due to the formation reaction kinetics, which is a process leaded by nucleation and growth of Al_2TiO_5 grains and finally the diffusion of the reactants remnants through the matrix, this is controlled for a very slow reacting species diffusion, as it was found by: Wohlfromm et al., (1991).

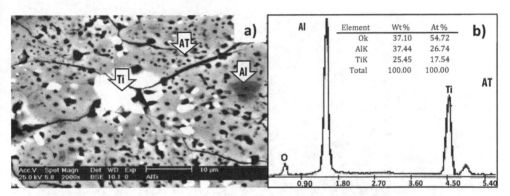

Fig. 10. a)BSE microstructure of Al_2O_3 and TiO_2 without addition, sintered at 1450°C for 3 hours, b) EDS of the matrix with an exact atomic relationship: 25 at% Al, 12.5 at% Ti and 62.5 at% O. (AT: Aluminum titanate; Ti: Titania; Al: Alumina).

The addition of the low melting point V_2O_5 (678°C) is evidenced in the microstructure with the presence of an abundant glassy intergranular phase, which constitutes a physical barrier between Al_2O_3 and TiO_2, retarding the Al_2TiO_5 formation (Fig. 11a).

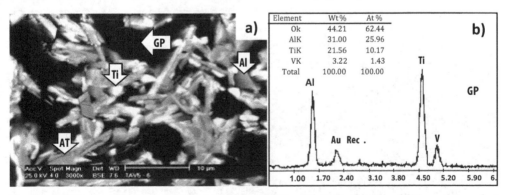

Fig. 11. a)BSE microstructure detail of Al_2O_3 and TiO_2 with V_2O_5: 6 wt% addition, sintered at 1450°C for 3 hours. b) EDS of the intergranular glassy phase (GP), appearing due to V_2O_5 low melting point. (AT: Aluminum titanate; Ti: Titania; Al: Alumina).

Althought localized EDS analysis on the glassy phase was carried out, the Al and Ti values obtained are due to the larger electron beam action volume compared to phase size (Fig. 11b).

The microstructure of MnO added samples shows extensive Al_2TiO_5 phase formation, with a minor presence of liquid phase, product of the two eutectic reactions, at 1290°C and 1330°C, between MnO and TiO_2. However, opposite to the V_2O_5 added samples, the reacting species diffusion and Al_2TiO_5 formation is accelerated with the MnO contents and, unreacted TiO_2 is absent in the microstructure, due to the secondary reactions. EDS analysis identified the intergranular eutectic phase as $2MnO.TiO_2$ and $MnO.TiO_2$ (Fig. 12). Microstructure grain size decreased with MnO contents in the sintered bodies.

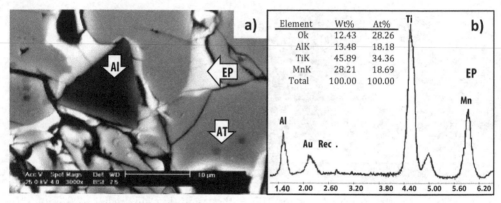

Fig. 12. a)BSE microstructural detail of Al_2O_3 and TiO_2 with MnO: 6 wt% addition, sintered at 1450°C for 3 hours. b) EDS of the intergranular $MnTiO_3$ eutectic phase (EP). (AT: Aluminum titanate; Al: Alumina).

The $FeSi_2.Si$ modified composition has a different microstructure to that obtained with other additives (Fig.13a-d).

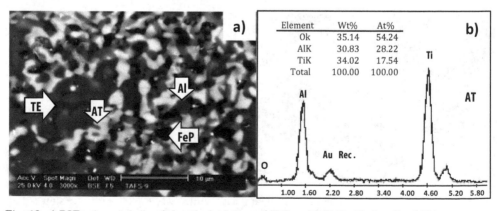

Fig. 13. a) BSE microstructural detail of Al_2O_3 and TiO_2 with $FeSi_2.Si$: 6 wt% addition, sintered at 1450°C for 3 hours. b)EDS of the Al2TiO5 matrix.

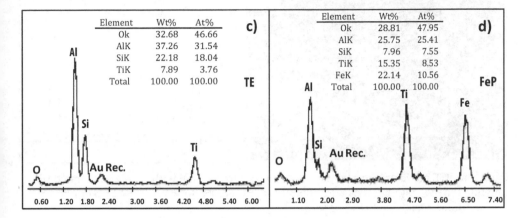

Fig. 13. c) EDS of ternary eutectic (TE) and d) EDS of the intergranular Fe rich phase (FeP).

A fraction of free Al_2O_3 remains as an intragranular phase, the TiO_2 reacts completely and besides the Al_2TiO_5 matrix phase a ternary eutectic reaction grainy phase is formed between the Al_2O_3-SiO_2-TiO_2 (as low as it is undetected by RXD), but increasing its quantity with the additive and, also a fourth intergranular phase rich in Fe is depicted (Fig.13d) (Arenas et al. 2011).

Pure and concentrated ilmenite ($FeTiO_3$), additions have a beneficial effect on grain growth control (Fig. 14 and 15). The SiO_2 left in the purified mineral promoted the formation of an intergranular liquid phase which could not be detected by XRD. Microstructures are practically free of unreacted original phases.

Fig. 14. a) BSE microstructural x1000 and a detail X3000, of Al_2O_3 and TiO_2 with pure $FeTiO_3$: 6 wt% addition, sintered at 1450°C for 3 hours. Notice the grain growth control.

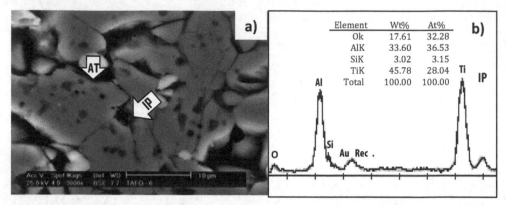

Fig. 15. a) BSE microstructural detail X3000, of Al_2O_3 and TiO_2 with with concentrated placer ilmenite ($FeTiO_3$. SiO_2): 6 wt% addition, sintered at 1450°C for 3 hours. b) EDS of intergranular phase due to SiO_2 presence.

6.4 Thermal stability

In order to determine the compositions stability, XRD analyses were performed on samples heat treated at 1100°C for 100 hours. The temperature selection is based on industrial applications working conditions and the maximum temperature for decomposition to occur.

The samples without additives and those with V_2O_5 and MnO showed a complete decomposition after heat treatment, as only the diffraction peaks of Al_2O_3 and TiO_2 were showed. (Fig. 16).

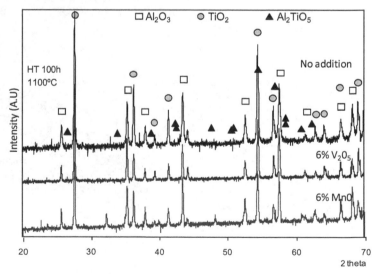

Fig. 16. X Ray Diffraction Al_2TiO_5 without addition, with 6wt% V_2O_5 and 6wt% MnO, heat treated for 100hours at 1100°C.

For FeSi$_2$.Si added samples (Fig.17), although the presence of oxidizing atmosphere, leads to the oxidation of Si and subsequent formation of ternary liquid phase between Al$_2$O$_3$, TiO$_2$ and SiO$_2$ which promotes a good densification, it has a minimal beneficial effect on stabilization. The presence of Al$_2$TiO$_5$ diffraction peaks is small if compared with that of Al$_2$O$_3$ and TiO$_2$, product of decomposition. It might be explained as, that only a fraction of the Fe ions from the FeSi$_2$ react and substitute the Al^{+3} ions, stabilizing the material.

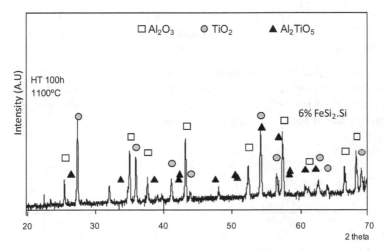

Fig. 17. X Ray Diffraction of Al$_2$TiO$_5$ with FeSi$_2$.Si 6 wt% addition, heat treated for 100hours at 1100°C.

The addition of pure FeTiO$_3$ (ilmenite) clearly shows an increase in the aluminum titanate stabilization (Fig. 18).

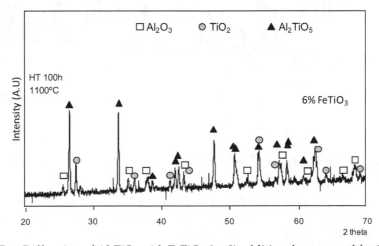

Fig. 18. X Ray Diffraction of Al$_2$TiO$_5$ with FeTiO$_3$ 6wt% addition, heat treated for 100hours at 1100°C.

This behavior agrees with the expected solid solution formation between FeTiO$_3$ and Al$_2$TiO$_5$, as depicted in the calorimetric studies (DSC). These experiments show the decomposition of ilmenite in air atmosphere to Fe$_2$O$_3$ and TiO$_2$ with the formation of Fe$_2$TiO$_5$ (Suresh et al.,1991), followed by Al$_2$TiO$_5$ reaction at higher temperature. This allows the possibility of a solid solution formation between Al$_2$TiO$_5$ and the isostructural Fe$_2$TiO$_5$ by a cation replacement mechanism.

From the ionic radii concept, the structural stabilization might be explained by the incorporation of Fe^{+3} (r=0.67Å) which decreases the structure distortion, caused by the Ti^{+4}: Al^{+3} radii difference (Shannon, R., 1969).

In the case of ilmenite addition, but from the concentrated mineral (Fig.19), the structural stabilization effect is evidenced only for the composition with 9% addition. This behavior might be attributed to the SiO$_2$ contamination which, on the other hand, benefits body densification by a liquid phase sintering mechanism.

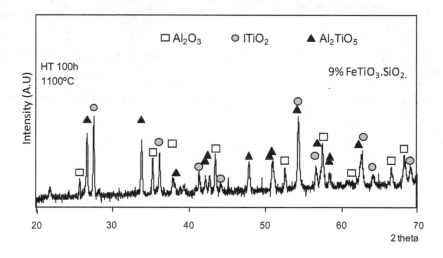

Fig. 19. X Ray Diffraction of Al$_2$TiO$_5$ with FeTiO$_3$.SiO$_2$ 9wt% addition, heat treated for 100hours at 1100°C.

The microstructure study by SEM-EDS corroborated the X-ray analysis evaluation (Fig. 20). All heat treated samples showed the characteristic elongated grain shape, typical of the Al$_2$TiO$_5$ decomposition into its original precursors raw materials Al$_2$O$_3$ and TiO$_2$. However, stabilized Al$_2$TiO$_5$ phase is observed in the compositions with 6% pure ilmenite, 9% concentrated mineral and in lower proportion in the samples with 6% ferrosilicon.

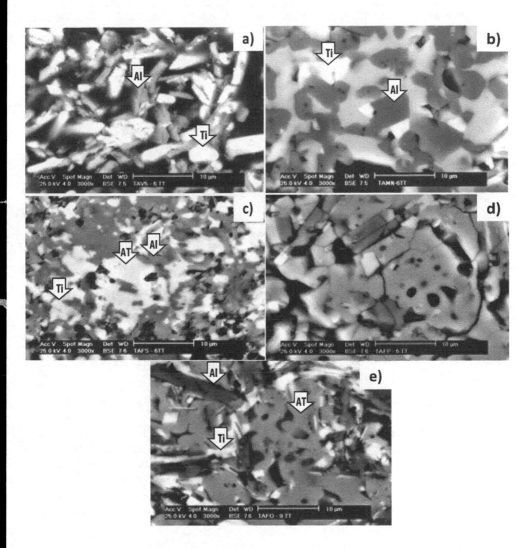

Fig. 20. a) BSE Microstructural detail x3000, Al_2TiO_5 with : a) V_2O_5 6wt% ; b) MnO 6wt%; c) $FeSi_2.SiO_2$ 6wt% d) $FeTiO_3$ 6wt% and e) $FeTiO_3.SiO_2$ 9wt% addition, heat treated at 1100°C for 100 hours. (AT: Aluminum titanate; Ti: Titania; Al: Alumina).

6.5 Al_2TiO_5 decomposition phase quantification

The quantification of Al_2TiO_5 decomposed after heat treatment was also determined using the internal standard method being the results showed in Table 3. The values obtained corroborate BSE image analysis, observations. The stabilization addition effect being higher in the compositions with pure and concentrated mineral, the MnO and the ferrosilicon stabilize slightly while vanadium oxide have not any effect.

AT+Additive (%)	%	%TiO_2 decomposed	% Al_2O_3 decomposed	%Al_2TiO_5 left
	3	43.86	55.82	0.32
V_2O_5	6	43.29	55.10	1.61
	9	43.71	55.63	0.66
	3	35.69	45.41	18.90
MnO	6	32.40	41.22	26.38
	9	32.90	41.84	25.26
	3	26.53	26.97	46.50
$FeTiO_3 .SiO_2$	6	23.06	29.30	47,64
	9	23.10	30.12	48.78
	3	17.17	21.78	61.05
$FeTiO_3$	6	10.28	13.01	76.71
	9	8.65	10.93	80.42
	3	38.01	48.36	13.63
$FeSi_2$	6	36.26	46.13	17.61
	9	33.20	42.24	24.56

Table 3. Al_2TiO_5 % Phase decomposition after heat treatment at 1100°C/ 100hours, by internal standard quantification method.

6.6 Grain size

Since the best stabilyzing behaviour after heat treatment was achieved by the ilmenites addition, these samples in the as sintered condition, were selected to determine the effect of additive contens on grains size. There were no significant variations in the grain size obtained with the two additives; however, there is a slight decrease in size as the percentage of the additive increases (in both cases), determined by image analysis of grain size (Table 4). The grain size varies between 9 and 12μm. In other words, the presence of second phase slightly inhibits the growth of grain.

$FeTiO_3$ (%)	Tg (μm)	$FeTiO_3.SiO_2$(%)	Tg (μm)
3	11.96	3	11.18
6	11.60	6	10.52
9	9.18	9	9,31

Table 4. Effect of $FeTiO_3$ (pure and concentrated mineral) on the grain size of sintered samples at 1450°C/ 3hours.

6.7 Mössbauer spectroscopy

It has confirmed the presence of the Fe^{+3} ions in all compositions with Fe added, i.e., the ilmenites and ferrosilicon.

The Mössbauer spectrum for material with 6% of ferrosilicon addition can be adjusted to two doublets (Fig.21a). The first doublet corresponds to the ferrous cation (Fe^{+2}) with a resonance that fits the hyperfine splitting with an isomer shift: IS = 1.01 ± 0.002 mm/s and a quadrupole splitting: QS= 0.664 ± 0.003 mm/s. The second doublet corresponds to the resonance of the ferric cation (Fe^{+3}) with a IS = 0. 323 ± 0.003 mm/s and a QS = 0. 520 ± 0.004 mm/s. For composition with 6% pure ilmenite addition (Fig. 21b.), it is revealed a consistent doublet with the ferric state (Fe^{+3}), with a IS = 0. 323 + 0.003 mm/s and QS = 0.520 ± 0.004 mm/s. The spectrum for the sample with 6% of mineral ilmenite (Fig. 19c), one could guess the doublet corresponds to both states ferrous (Fe^{+2}) and ferric (Fe^{+3}), however should be noted that results are not accurate, with considerable dispersion.

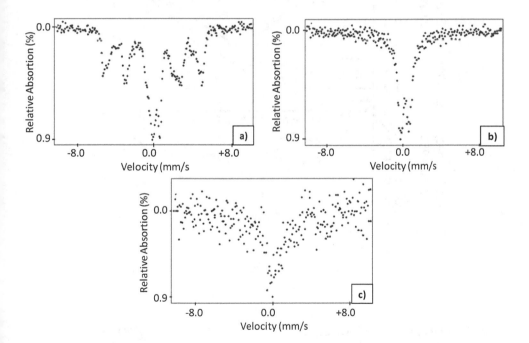

Fig. 21. Mössbauer Spectroscopy of Al_2TiO_5 with: a) 6wt% $FeSi_2.SiO_2$; b) 6wt% $FeTiO_3$ and c) 6wt% $FeTiO_3.SiO_2$ additions.

These results corroborate the possible replacement of the Al^{+3} ions by ion Fe^{+3}; there is higher stabilization in samples with pure ilmenite addition, where all Fe ions are in ferric state. However, as it was found in previous research (Barrios de Arenas & Cho, 2010) , the presence of Fe^{+2} ions, also represent the possibility of Al^{+3} ions substitution, with the creation of defects, which in turn promotes the diffusion in solid state.

6.8 Thermal expansion

Some authors have directly related the area of hysteresis in thermal expansion curves to sample microcracks density (Lingenberg W. 1985). In this study, after comparison of the ilmenite added and pure Al_2TiO_5 materials results (Fig.22), show an evident reduction in the area of hysteresis in the formers, being even more important in the samples with concentrated mineral $FeTiO_3$ addition (fig.22 b).

In both $FeTiO_3$-added samples, the property values are antagonist to addition contents.

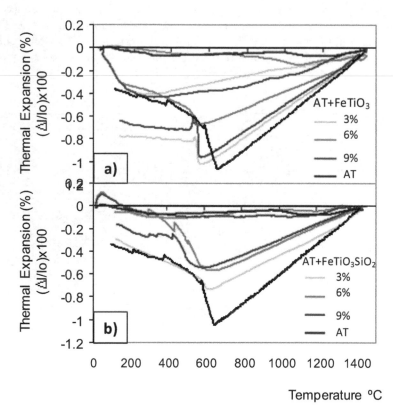

Fig. 22. Thermal Expansion betweeen 20 -1450°C of Al_2TiO_5 with a) $FeTiO_3$: 6wt% and b) $FeTiO_3.SiO_2$ 6wt% addition.

Table 5, lists the experimental values obtained for the thermal expansion coefficients between 25-1000°C and 25-1450°C. The Al_2TiO_5 without additives has a value slightly negative $\alpha_{25-1000°C}= -0.55 \times 10^{-6}°C^{-1}$. Both additions turn this characteristic value into very low positive ones augmenting with the additive content in each case. This behavior could be explained by the reduction in grain size which causes a microcracking inhibition, as grains boundaries surface increases, i.e., it is necessary a higher energy for cracking (Yoleva et al. 2010).

Composition	$\alpha_{25-1000°C}$ x10^{-6} °C^{-1}	$\alpha_{25-1450°C}$ x10^{-6} °C^{-1}
Al$_2$TiO$_5$	-0.55	0.87
FeTiO$_3$ (%)		
3	0.62	0.96
6	0.76	1.05
9	0.86	1.16
FeTiO$_3$.SiO$_2$ (%)		
3	0.83	1.11
6	0.94	1.22
9	1.02	1.40

Table 5. Effect of FeTiO$_3$ (pure and concentrated mineral) addition, on Al$_2$TiO$_5$ Thermal Expansion.

7. Conclusion

The evident effect of FeTiO$_3$ additions on Al$_2$TiO$_5$ was established. An increase of the addition leads to a sensitive decrease on decomposition, this is due to the expected solid solution formed between Al$_2$TiO$_5$ and the isostructural Fe$_2$TiO$_5$, the latter being the product of ilmenite decomposition in the oxidizing conditions used.

The ilmenites additions produce a slightly increase in the thermal expansion coefficients being more important with the concentrated mineral addition. Although the values remain acceptable.

The MnO increases densification by the presence of a localized liquid phase, which allows the rearrangement of particles in the first stage of sintering and also slightly decomposition controls.The ferrosilicon (FeSi$_2$.SiO$_2$), SiO$_2$ reacts with Al$_2$O$_3$ and the TiO$_2$ forming a liquid phase, allowing rearrangement of particles; however, it lacks of stabilizing effect.

8. References

Asbrink, G. & Magneli, A. (1967) "X-Ray Studies on Some Mixed Oxides Systems of Pseudobrookite Structure", *Acta Chem. Scand* 21.

Bachmann, J. L. (1948) "Investigations of Properties of Aluminium Oxide and Some Aluminuos Materials"; *Ph.D Thesis, Pennsylvania State University.*

Barrios de Arenas I. & Cho S–A. (2011) "Efecto de la adición de ferrosilicio - FeSi$_2$ en la microestructura y estabilidad del titanato de aluminio- Al$_2$TiO$_5$." *Revista Latinoamericana de Metalurgia y Materiales* 2011; 31 (1): 11-19.

Barrios de Arenas I. (2010) "Estudio de materiales Compuestos de titanato de aluminio Al_2TiO_5 estabilizado con ilmenita Fe_2TiO_5, reforzado con Al_2O_3 y TiO_2", *Revista Latinoamericana de Metalurgia y Materiales*; 30 (2): 210-22.

Bayer, G. (1971) "Thermal Expansión Characteristics and Stability of Pseudobrookite Compounds, Me_3O_5". *Less Common Metals*, 24,[3],129-138.

Bayer, G. (1973) "Thermal Expansion Anisotropy of Oxide Compounds", *Proc. Brit. Ceram. Soc.* 22, de. D. J. Godfrey. Brit.Cer.Soc., Stoke on Trent, 39-53.

Brown, I. & Mc Gavin, D. (1994) "Effect of Iron Oxides Additives on Al_2TiO_5 Formation" Fourth Euroceramics. Vol 4. 487-492. Faenza editors. Italy

Buessem, W. R .& Lange, F. F.: "Residual Stresses in Anisotropic Ceramics", *Interceram.* 15 [3] (1966) 229-231.

Buscaglia, V.; Alvazzi, M.; Nanni, P.; Leoni, M. & Bottino, C. (1995) "Factors Affecting Microstucture Evolution During Reaction Sintering of Al_2TiO_5 Ceramics" Ceramics Charting The Future. Edit. P. Vicenzini, Techna Srl. 1867-1875.

Buscaglia, V.; Carracciolo, F.; Leoni, M.; Nanni, P.; Viviani, M. & Lematre, J. (1997) "Synthesis, Sintering and Expansion of $Al_{0.8}Mg_{0.6}Ti_{2.1}O_5$ Low Thermal Expansion Material Resistent to Thermal Decomposition" *J. Mater. Sci.* 32 6525-6531.

Buscaglia, V.; Musenich R.; Nanni, P. & Leoni, M. (1996) "Solid State Reactions in Ceramics Systems" pp. 123-26 in Proceedings of International Workshop on Advanced Ceramics´96, March 12-14, Inuyama, Japan.

Cleveland J. J. & Bradt R. C. (1978) "Grain Size Microcracking Relation for Pseudobrookite Oxides", *J. Am. Ceram. Soc.* 61 [11-12] 478-481.

Cleveland, J.J. (1977) Critical Grain Size for Microcraking in the Pseudobrookite Structure; M.S. Thesis, The Pennsylvania State University, March.

Epicier, T.; Thomas, G.; Wohlfromm, H.; Moya, J. (1991) "High Resolution Electron Microscopy of the Cationic Disorder in Al_2TiO_5". *J. Mater. Res.* 6 [1] 138-145.

Fonseca, A. P & Baptista, J. (2003) "Efecto de la estequiometria y la temperatura de cocción en el desarrollo de la fase Al_2TiO_5". Bol. Soc. Esp. Ceram. Vidrio. 42 [2] 65-68.

Freudenberg, B.: "Etude de la reaction à l'état solide: Al_2O_3 + TiO_2 - Al_2TiO_5", Tesis Doctoral, Eĉole Polytécnique, Lausanne 1987.

Golberg, D. (1968) "On the Systems Formed by Alumina with Several Trivalent and Tetravalent Metal Oxides, in particular Titanium Oxide" *Rev. Int. Hautes Temp. Refract.*, 5, 181-94.

Hasselman, D. P. & Singh, J. P. (1979) "Analysis of Thermal Stress Resistance of Microcracked Brittle Ceramics"; *Bull. Am. Cer. Soc.* 58, [9] 856-860.

Holcombe, C. E. & Coffey, A .L. (1973) "Calculated X-Ray Powder Diffraction Data for β-Al_2TiO_5"; *J. Am. Ceram. Soc.* 56 [4] 220 -221.

Huber J. & Heinrich, J. (1988) Keramic im Motor; en ENVICERAM ´88, 1st. Int. Symp. And Exposition of Ceramics for Environmental Protection, 7-9 Dic. 1988, Köln, FRG.

Ishitsuka, M., Sao, T., Endo, T. & Shimada, M. (1987) "Synthesis and Thermal Stability of Aluminium Titanate Solid Solutions" *J. Am. Ceram. Soc.*, 70 69-71.

Jung, J.; Feltz, A.& Freudenberg, B. (1993) "Improved Thermal Stability of Al – Titanate Solid Solutions", *Cfi/ Ber. DKG* 70 No. 6 299-301.

Jürgen, H. & Aldinger, F. (1996) "Applied Phase Studies", *Z. Metallkd.* 87[11] 841-848.

Kato, E., Daimon, K. & Takahashi, J. (1980) "Decomposition Temperature of β- Al_2TiO_5", *J. Am. Ceram. Soc.* 63 (1980) 355.

Klug, H. & Alexander, L. (1954) "X Ray Diffraction Procedures" Edit. John Wiley, p. 410, New York.

Kolomietsev, V.; Suvorov, S.; Makarov, V. & Butalov, S.(1981) "Sintering and Some Problems of Composite Materials in the Al_2O_3 -Al_2TiO_5 System" *Refractories* 22, 627-631.

Kuszyk J. A. & Bradt, R. C.(1973) "Influence of Grain Size on Effects of Thermal Expansion Anisotropy in $MgTi_2O_5$", *J. Am. Ceram. Soc.*, 56 [8] 420-23.

Lang, S.; Fillmore, C. & Maxwell, L. (1952) "The System Berillia-Alumina-Titania: Phase Relations and General Physical Properties of Three Components Porcelains", *J. Res. Natl. Bur. Stand.*, 48 [4] 301-321.

Lingenberg, W. (1985) "Werkstoffeigenschaften von Al_2TiO_5 unter besonderer Berücksichtigung von Bildungs und Zerfallsreaktionen; Disertation, TU Clausthal.

Liu, Z.; Zhao, O. &Yuan J. J. (1996) "The Effects of Additives on Properties a Structure of Hot Pressed Aluminium Titanate Ceramics". *J. Mater. Sci. Lett.* 31 90-94.

Milosevski, M. (1997) "Thermal Diffusivity of Al_2TiO_5, $CaTiO_3$ and $BaTiO_3$" *Science of Sintering* 29 [2] 105-112.

Milosevski, M.; Ondracek, O.; Milisevska, R.; Spaseska, D. & Dimeska, A. (1995) "Thermal Expansion and Mechanical Properties of Al_2TiO_5 – SiO_2 System" Ceramics Charting The Future. Edit. P. Vicenzini, Techna Srl. 1875-1882.

Morishima, H.; Kato, Z.; Uematsu, K.; Saito, K.; Yano, T.; Ootsuka, N. (1987) "Synthesis of Aluminium Titanate - Mullite Composite having High Thermal Shock Resistance"; *J. Mater. Sci. Lett.* 6 389-390.

Morosin, B. & Lynch, R. W. (1972) "Structure Studies on Al_2TiO_5 at Room Temperature and 600°C"; *Acta Crystallogr.* Sect. B 28 1040.

Navrostky, A. (1975) "Thermodynamics of Formation of Some Compounds with the Pseudobrookite Structure and of the $FeTi_2O_5$-Ti_3O_5 So lid-Solutions Series"; *American Mineralogist* 60 249-256.

Ohya, Y. & Nakagawa, Z. (1996) "Measurement of crack volume due to thermal expasión anisotropy in aluminium titanate ceramics", *J. Mater. Sci.* 31 1555-1559.

ShannoN, R.D. & Pask, J.A. (1965) "Kinetics of the Anatase-Rutile Transformation"; *J. Am. Ceram. Soc.* 48 391-398.

Sheppard L. M. (1989) "A Global Perspective of Advanced Ceramics", *Am. Ceram. Soc. Bull.* 68 (9) 1624-1633.

Sleepetys, R. & Vaughan, P. (1969) "Solid Solution of Aluminium Oxide in Rutile Titanium Dioxide" *J. Phys. Chem.*, 73, , 2157 - 62.

Stingl, P.; Heinrich, J. & Huber, J. (1986) Proceedings of the 2nd International Symposium of ceramic Materials and Component Engines, Lübeck - *Travemunde (FRG)* April 1986. Edited by W. Bunk and H. Hausner. DKG Bad Honned 369.

Thomas, H. & Stevens, R. (1989) "Alumina Titanate - A Literature review. Part. 2 Engineering Properties and Thermal Stability"; *Br. Cer. Trans. J.* 88 184-190.

Tilloca, G. (1991) "Thermal Stabilization of Aluminium Titanate and Properties of Aluminium Solid Solutions" *J. Mater. Sci.* 26 2809-2814.

Woermann, E. (1985) Die Thermische Stabilität von Pseudobrookit-Mischkristallen; DFG-Abschlu-bericht Tialit (AZ. W. 81/23).

Wohlfromm, H.; Moya T. S. & Pena P. (1990) "Effect of $ZrSiO_4$ and MgO Additions on Reaction Sintering and Properties of Al_2TiO_5 Based Materials". *J. Mater. Sci* 25 3753-3764.

Yoleva, A.; Djambazov, S.; Arsenov, D. & Hristov, V. (2010) "Effect of SiO_2 addition on thermal hysteresis of Aluminum titanate" J. Univ. of Chem. Tech. and Met. 45 3 269-274.

7

Effects of the Microstructure Induced by Sintering on the Dielectric Properties of Alumina

Zarbout Kamel[1], Moya Gérard[2],
Si Ahmed Abderrahmane[2], Damamme Gilles[3] and Kallel Ali[1]
[1]Sfax University, LaMaCoP, BP 1171, Sfax 3000,
[2]Aix-Marseille University, Im2np, UMR-CNRS 6242, Marseille,
[3]Commissariat à l'Energie Atomique, DAM Ile-de-France, Bruyère-le-Châtel
[1]Tunisie
[2,3]France

1. Introduction

The dielectric properties of undoped corundum, α-alumina, of different kinds (single crystal or polycrystalline obtained by solid state sintering) have been the subject of numerous studies concerning, in particular, the breakdown strength and the charging behavior (Haddour et al., 2009; Liebault et al., 2001; Si Ahmed et al., 2005; Suharyanto et al., 2006; Thome et al., 2004; Touzin et al., 2010; Zarbout et al., 2008, 2010). The common feature pointed out by most of these investigations is the conspicuous role played by the microstructure and the impurities. It is also established that the microstructure induced by the sintering process (grain size and porosity) goes concomitantly along with impurities segregation at grain boundaries and/or the development of defects in the lattice (Chiang et al., 1996; Gavrilov et al., 1999; Lagerlöf & Grimes, 1998; Moya et al., 2003). To some extent, for a given composition, these evolutions can be governed by the sintering conditions, for instance the firing cycle in the case of solid state sintering (Chiang et al., 1996).

The breakdown strength is a key parameter for the reliability of dielectrics and in particular of microelectronic insulator components. This parameter is intimately linked to the charging properties as breakdown originates from the enhancement of the density of trapped charges, which stems from the competition between charge trapping and conduction (Blaise & Le Gressus, 1991; Le Gressus et al., 1991; Liebault et al., 2001; Haddour et al., 2009). The charges can be either generated by irradiation or injected through interfaces via an applied voltage. Charge trapping can occur around intrinsic point defects, defects induced by the dissolution of impurities, defects associated with grain boundaries interfaces and dislocations (Kolk & Heasell, 1980). We must also keep in mind that trapping in insulators gives rise to polarization and lattice deformation allowing energy accumulation within the material (Blaise & Le Gressus, 1991; Le Gressus et al., 1991; Stoneham, 1997). As a result, if some critical density of trapped charges (or some critical electric field) is reached, external stresses (thermal, electrical or mechanical) can trigger a collective relaxation process corresponding to a release of stored energy. If the amount of this energy is sufficient, breakdown could set in causing irreversible damages of the material (Blaise & Le Gressus, 1991; Moya & Blaise, 1998; Stoneham, 1997).

It appears that an improvement of the breakdown strength would require that conduction, which tends to decrease density of trapped charges, be favored to some extent without, however, substantially altering the insulating properties. Conduction will also be referred as the ability of the material to spread charges. Therefore, the control of the competition between charge accumulation (trapping) and spreading (conduction), via the fabrication processes, is a key technological concern. The foregoing arguments motivate the need to develop methods for the characterization of charge conduction (conversely charge trapping) and underscore furthermore the importance of controlling the microstructural development during sintering of ceramic insulators. The purpose of this chapter is to provide the physical background for a more comprehensive understanding of the effects of the microstructure (and the various defects) induced by the sintering conditions on charge conduction in alumina. This understanding, which could be generalized to other ceramics, appears as prerequisite for the fabrication of insulators of improved dielectric breakdown strength.

2. Defects in α-alumina

The α-alumina, exhibits the hexagonal corundum structure. In this structure, Al³⁺ cations occupy only two-thirds of the available sites and an interstitial unoccupied site arises between alternate pairs of Al³⁺. Charge trapping in alumina may take place around defects that can be intrinsic in nature or stemming from the dissolution of impurities (i.e., the foreign cations and their charge compensating defects). In sintered materials, one has also to take into account the effect of grain boundaries, segregation of impurities and defects at interfaces. These defects are characterized by energy levels within the wide band gap (of about 9 eV in alumina).

2.1 Point defects

2.1.1 Intrinsic point defects (Schottky and Frenkel defects)

The Schottky defects consist of pairs of negatively charged cationic vacancies $V_{Al}^{'''}$ and positively charged anionic vacancies $V_O^{\bullet\bullet}$. The vacancies must be formed in the stoichiometric ratio (two aluminium for three oxygen) in order to preserve the electrical neutrality of the crystal. Using the Kröger-Vink notation, the formation of Schottky defects obey to the reaction:

$$2\,Al_{Al}^x + 3\,O_O^x \Leftrightarrow 2\,V_{Al}^{'''} + 3\,V_O^{\bullet\bullet} + Al_2O_3 \tag{1}$$

The Frenkel defects are formed when the Al³⁺ cation (Eq. 2) or the O²⁻ anion (Eq. 3) is displaced from its normal site onto an interstitial site giving a vacancy and an interstitial pair:

$$Al_{Al}^x \Leftrightarrow V_{Al}^{'''} + Al_i^{\bullet\bullet\bullet} \tag{2}$$

$$O_O^x \Leftrightarrow V_O^{\bullet\bullet} + O_i^{''} \qquad (3)$$

Simulation results show that the formation energies of intrinsic point defects in α-alumina are relatively high (Atkinson et al., 2003). They are estimated respectively for Schottky defects, cation Frenkel and anion Frenkel at 5.15, 5.54 and 7.22 eV.

2.1.2 Extrinsic point defects

Extrinsic point defects are entailed by the dissolution of foreign elements. The solubility of an impurity depends mainly on its cation size (generally, small size elements exhibit high solubility). The charge compensating defects accompanying the dissolution of aliovalent impurities (i.e, defects that are required for ensuring the electrical neutrality) are determined not only by their valence (charge) but also by their position (interstitial or substitutional) in the host lattice.

In the case of a cation (M) greater in valence than the host cation (Al³⁺), the dissolution mode in substitutional position is most likely the cationic vacancy compensation mechanism (Atkinson et al., 2003). Accordingly, for tetravalent cations, in MO_2 (such as SiO_2 or TiO_2), the dissolution reaction is:

$$3\,MO_2 + 4\,Al_{Al}^x \Leftrightarrow 3\,M_{Al}^{\bullet} + V_{Al}^{'''} + 2\,Al_2O_3 \qquad (4)$$

This compensation by a cationic vacancy is somewhat corroborated by experiments involving solution of Ti⁴⁺ in α-Al₂O₃ (Mohapatra & Kröger, 1977; Rasmussen & Kingery, 1970).

For divalent cations, in MO (such as MgO or CaO), the anionic vacancy compensation of substitutional $M_{Al}^{'}$, is suggested (Atkinson et al., 2003):

$$2\,MO + 2\,Al_{Al}^x + O_O^x \Leftrightarrow 2\,M_{Al}^{'} + V_O^{\bullet\bullet} + Al_2O_3 \qquad (5)$$

The interstitial dissolution of monovalent elements, in M_2O (such as Na_2O or Ag_2O), can be governed by a host cationic vacancy compensation mechanism. However, a self-compensating dissolution mode, involving both interstitial and substitutional positions of M, is also expected (Gontier-Moya et al., 2001).

In the case of isovalent elements, in M_2O_3 (such as Cr_2O_3 or Y_2O_3), the dissolution will not create charged defects in the lattice but can induce a stress field due to the misfit arising from the difference in cation radii.

As previously pointed out, the formation energies of intrinsic defects are very high in α-alumina. Consequently, a few ppm of impurities will make the concentrations of extrinsic defects higher than those of the intrinsic ones, even at temperatures near the melting point (Kröger, 1984; Lagerlöf & Grimes, 1998).

2.1.3 Point defects association

Isolated point defects can be associated, at appropriate temperature, to form neutral or charged defect clusters. This association leads to a substantial reduction in the solution energy due to strong coulombic interaction and lattice relaxation. A typical example of defect clustering is the association of defects induced by the dissolution of tetravalent cations (Eq. 4):

$$3\,M^{\bullet}_{Al} + V'''_{Al} \Leftrightarrow (3\,M^{\bullet}_{Al} : V'''_{Al})^{x} \qquad\qquad (6)$$

Mass action calculations (Lagerlöf & Grimes, 1998) have shown that the relative concentrations of extrinsic defects (point defects and defect associations) depend on the equilibrium temperature under which they are created. In sintered alumina, they can be determined by the sintering temperature and time, i.e. isothermal part of the firing schedule.

2.1.4 Association of point defects with charges

Anionic vacancies $V^{\bullet\bullet}_O$ (or cationic vacancies V'''_{Al}) can be associated with electrons (or holes) to form F centers (or V centers). In fact, upon trapping one electron (or two electrons), $V^{\bullet\bullet}_O$ becomes a F$^+$ center (or a F center). Anionic vacancies can also be associated with a substitutional divalent cation M'_{Al} to form F$_{cation}$ center (such as F$_{Mg}$ or F$_{Ca}$). The F, F$^+$ and F$_{cation}$ act as a donor centers. The energy levels of F and F$^+$ are respectively estimated around 3 and 3.8 eV below the edge of the conduction band (Kröger, 1984).

2.2 Extended defects: Grain boundaries

Grain boundaries are the interfaces between like crystals, at which atomic planes are always disrupted to some extent. The atomic order of the lattice is preserved up to within approximately a unit cell of the dividing plane. Thus, the disordered region of a grain boundary is typically only 0.5 − 1 nm wide, although it does vary somewhat with the type of boundary and the crystal lattice periodicity. Grain boundaries provide segregation sites for impurities and defects.

3. Sintering of α-alumina

3.1 Sintering of α-alumina and induced microstructure

The polycrystalline alumina samples (of 0.2 cm thickness and 1.6 cm diameter) were processed, at Ecole Nationale Supérieure des Mines "ENSM" of Saint Etienne (France), by sintering from two types of powders of different purities (Table 1). The first referred below as "pure", elaborated by the Exal process and provided by CRICERAM Co., contains about 150 ppm of different impurities (with 90 ppm of silicon). The second referred as "impure", elaborated by the Bayer process and provided by REYNOLDS Co., has an impurity content near 4000 ppm (with 1497 ppm of silicon). In Table 1, the compositions of single crystals, which will be considered as reference material, are also given (cf. next section).

Before sintering, the powders were prepared according to conventional procedures involving successively aqueous dispersion, adding of organic binders, spray drying, uniaxial die forming and cold isostatic pressing. Sintering near theoretical density was performed in air with a firing schedule (Fig. 1), which comprises:

- a binder burnout stage where samples were heated slowly (1 K/min) up to 773 K in three steps during 5 hours,
- a heat-up stage, at 2 K/min, from 773 K to the required sintering temperature T_s,
- an isothermal heat treatment at T_s with a dwelling time t_s,
- a cooling stage at rate of about 10 K/min.

	SiO$_2$	CaO	MgO	Na$_2$O	Fe$_2$O$_3$	K$_2$O	Cr$_2$O$_3$	TiO$_2$
Alumina powders "pure" (Criceram)	90	5	< 5	40	12	---	---	---
Alumina powders "impure" (Reynolds)	1497	686	723	404	415	---	---	---
Single crystal (RSA)	290	16	< 10	19	48	---	---	---
Single crystal (Pi-Kem)	---	0.6	0.2	1.5	9	2.5	0.4	0.3

Table 1. Composition of alumina materials (impurities in ppm).

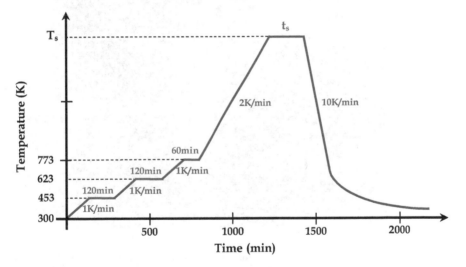

Fig. 1. Schematic description of the firing schedule of the sintering process.

The grain diameters, d, and densities of the sintered samples, which were achieved via the control of the sintering temperature, T_s, and the dwelling time, t_s, are given in Table 2.

The grain sizes were determined by the intercept method. The sample surfaces were first polished and then thermally etched to reveal the grain boundaries. Etching was performed by holding the sample at 50 K below the sintering temperature T_s (during 15 to 30 min) after a rapid heating. The average grain sizes were calculated from the values of 100 to 200 grain size measurements from Scanning Electron Microscope "SEM" images of the surface using different magnifications. Figure 2 shows, as example, the microstructure of the 1.2 μm grain diameter of the impure polycrystalline alumina sample.

	grain diameter, d (μm)	sintering temperature, T_s (K)	dwelling time, t_s (min)	density (in % of theoretical)
"pure" samples	1.7	1863	100	98.7
	2.7	1893	120	98.9
	4.5	1923	120	99.1
"impure" samples	1.2	1773	180	95.7
	2	1823	180	98
	4	1923	360	95.4

Table 2. Sintering temperature, dwelling time at the sintering temperature and corresponding grain diameters and densities (Liebault, 1999).

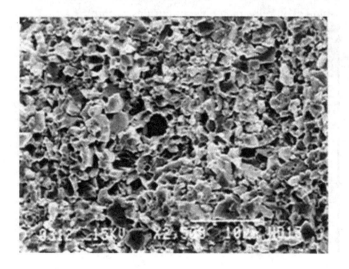

Fig. 2. Microstructure of the 1.2 μm grain diameter of the impure polycrystalline alumina sample (SEM image).

3.2 Single crystals (as reference materials)

The single crystals are selected for the purpose of providing reference materials, which will be compared to the sintered alumina. Two types of single crystals taken from a Verneuil-grown rod are considered (Table 1). The first of very low impurity content (about 15 ppm) is provided by Pi-Kem Co. (U.K.). The second, manufactured by RSA Co. (France), contains about 380 ppm of various impurities (with 290 ppm of silicon). The samples were polished to flat mirror surface finishes, using successively finer grades of diamond pastes down to 1 μm. In order to anneal the defects induced by machining and polishing, a thermal treatment in air at 1773 K during 4 hours was performed.

3.3 Characterization of defects in sintered alumina using positron annihilation lifetime spectroscopy

In this section, complementary investigations concerning the use of Positron Annihilation Lifetime Spectroscopy "PALS" for the characterization of defects in alumina will be summarized (Moya et al., 2003; Si Ahmed et al., 2005). After a short description of the procedure, we will spotlight the results relevant to the charging properties.

3.3.1 Experimental procedure

The positron lifetime spectra of the samples, identical to those described in Table 1, were recorded at room temperature using a conventional fast-fast coincidence system with γ detectors consisting of plastic scintillators characterized by a time resolution of 270 ps. The ^{22}Na positron source, 10 μCi sealed by 0.75 μm thick nickel foils, was sandwiched between two identical samples (of 20 mm diameter and 2 mm thickness). The spectra were measured in 2000 channels, with a calibration of 27 ps/channel, collecting 6.9×10^6 to 20×10^6 counts.

3.3.2 Spectra analysis

The experimental spectra were fitted via a LTV.9 program (Kansy, 1996), in which a three-state trapping model was introduced into the source code (Krause-Rehberg & Leipner, 1999).

In the case of single crystal Pi-Kem, the spectra analysis reduces to only one lifetime component (τ_b = 117 ± 1 ps) associated with reasonable confidence to annihilation in bulk free defects. The quite high intensity of this component, (98.4 ± 1.6) %, appears as an unequivocal justification of the absence of discernable defects that are able to trap positrons (e.g., dislocations, negatively charged vacancies, neutral complexes). This result is also interpreted as a confirmation of the very low impurity content, which therefore justifies its choice as reference material for assessing the effects of both the microstructure achieved by the sintering process and the impurity content. For the sintered samples, three lifetime components were deduced. The shortest lifetime (τ_b = 122 ± 4 ps) was attributed to annihilation in the bulk free defects as it is very close to the one of the reference material, the intermediate (τ_g = 137 ± 2 ps) to trapping in defects within the grains and the longest (τ_{gb} = 397 ± 22 ps) to trapping in clusters located at grain boundaries.

Since Silicon is the main impurity in sintered samples, the possible defects felt by positrons within the grains are isolated vacancies $V_{Al}^{'''}$ and $(Si_{Al}^{\bullet}:V_{Al}^{'''})''$, $(2Si_{Al}^{\bullet}:V_{Al}^{'''})'$, $(3Si_{Al}^{\bullet}:V_{Al}^{'''})^x$ clusters. However, the neutral cluster $(3Si_{Al}^{\bullet}:V_{Al}^{'''})^x$ is more likely than the others due to the sintering temperatures and dwelling times (Lagerlöf & Grimes, 1998). In any case, the positron lifetime in all these defects is expected to have about the same value (τ_g = 137 ± 2 ps) because it is primarily determined by $V_{Al}^{'''}$ (Moya et al., 2003). The nature of clusters at grain boundaries is more difficult to ascertain because of the competition between all impurities for the segregation sites. However, one can speculate that the lifetime of about 400 ps (τ_{gb} = 397 ± 22 ps) reflects positron trapping in neutral or negatively charged clusters of charge compensating native vacancies $V_{Al}^{'''}$ and $V_O^{\bullet\bullet}$, which are induced by the impurities that have the strongest tendency for segregation, for instance SiO_2, MgO and CaO. In particular, $V_O^{\bullet\bullet}$ could stem from the dissolution of CaO (Eq. 5), which, incidentally, displays by far the highest enrichment ratio, about 1300 (Dörre & Hübner, 1984).

4. Method for the characterization of the charging state of dielectrics

In insulator material, the charges can be injected through interfaces, via an applied voltage, or created by irradiation via energetic particles. For instance, incident electrons, as they slowdown, can generate pairs of electrons and holes. These charge carriers can recombine, be trapped or move as a result of diffusion and/or field conduction. Concurrently, some of the electrons can be emitted from the sample surface and a distribution of trapped charges may develop within the irradiated volume. In our case, experiments are carried out using the electron beam of a SEM. The experimental set up, described below, provides means for measuring the evolution of the net amount of trapped charges Q_t, which characterize the charging state of the insulator. For this purpose, we use the Induced Current Measurements "ICM" method (called also the Displacement Current Measurements), which we have recently improved (Zarbout et al., 2008, 2010).

4.1 Experimental set up

The experiments are performed using a SEM (LEO 440), which is specially equipped (Fig. 3) to inject a controlled amount of charges Q_{inj} with appropriate conditions (energy of incident electrons, charge and current beam densities, temperature). The electron beam is monitored by a computer system allowing the control of the beam parameters:

-    beam energy, E_p (varying between 300 and 40000 eV),
-    current beam intensity, I_p (ranging from a few pA to several hundred nA, with the possibility of reaching a few µA),
-    irradiated beam area of the sample (varying between a few nm when the beam is focused and few hundred µm when it is defocused).

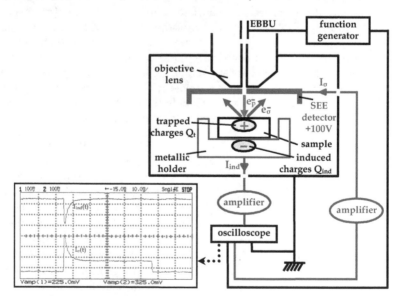

Fig. 3. Setting up for the measurement of the secondary electron current I_o and induced current I_{ind}. The secondary electron detector (diameter 15 cm) is positioned at a distance of about 2 cm from the sample surface.

Furthermore, the injection time, t_{inj} (ranging from 10^{-3} to 1 s), is adjusted by the Electron Beam Blanking Unit "EBBU" using a function generator, allowing the turns off on the spot over the specimen surface with no electron track outside the investigated area when the beam is blanked.

The metallic sample holder is attached to a cooling-heating stage (temperature range 93 – 673 K). Hence, in situ thermal sample cleaning under vacuum (at T = 663 K during 180 minutes) and sample characterization at different temperatures are possible. Prior to electron irradiations, after the cooling that follows the cleaning step, the sample is held during 180 minutes at the testing temperature, so that the thermal equilibrium between the sample and the metallic holder is approached.

4.2 The improved Induced Current Measurement method

The ICM method is based on the measurement of the current I_{ind}, produced by the variation of the induced charges Q_{ind} (in the sample holder) due to the trapped charges in the sample Q_t (Liebault et al., 2001, 2003; Song et al., 1996). Since the influence coefficient in our experimental set up is close to one (Zarbout et al., 2008), the amount of the net trapped charges is given by:

$$Q_t(t) = -Q_{ind}(t) = -\int_0^t I_{ind}(t)\,dt \qquad (7)$$

The improvement brought to the ICM method consists in the concurrent measurement of I_{ind} and the total secondary electron current I_o due to the sole electrons emitted by the sample. For this purpose, as shown in Fig. 3, the SEM is specially equipped with a secondary electron low-noise collector located under the objective lens just above the sample. A biased voltage of 100 V is applied to it in order to collect all the electrons escaping from the sample surface.

During charge injection, the currents, I_o and I_{ind}, are simultaneously amplified (Keithley 428) and observed on an oscilloscope (HP 54600B) where the material response is displayed after a short lag time. The primary current beam I_p (which is adjusted in a Faraday cage) and the current I_o are always positive whereas I_{ind} can be positive or negative depending on the Secondary Electron Emission "SEE" yield σ, which is equal to I_o/I_p. Then if I_o is higher than I_p ($\sigma > 1$), the sample charges positively and I_{ind} is negative. In the other case ($\sigma < 1$), the sample charges negatively and I_{ind} is positive. The general variation of σ with E_p is shown in Fig. 4.

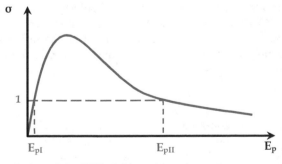

Fig. 4. Schematic evolution of the SEE yield σ with primary beam energy E_p for uncharged insulator materials.

The sign of the net sample charges (i.e., net trapped charges) depends mainly, for a given material, on the primary beam energy E_p and the primary current beam density J. In the case of α-alumina, a sign inversion of the net trapped charges is observed for J greater than 7×10^6 pA/cm^2 (Thome et al., 2004).

Experiments are performed with a 10 keV primary electron beam energy which is located, for the α-alumina materials, between the two crossover energies of primary electrons E_{pI} and E_{pII} (about 20 keV) for which σ is equal to 1. To probe a zone representative of the material we use a defocused beam over an area of 560 μm diameter Φ, which has been accurately measured using an electron resist (Zarbout et al., 2005). With the utilized value I_p = 100 pA, the primary current beam density is J = 4×10^4 pA/cm^2. These experimental conditions give rise to net positive trapped charges (σ > 1) and then a negative induced current, as shown in the curves recorded by the oscilloscope in Fig. 3.

The positive surface potential developed by the trapped charges Q_t does not exceed a few volts, which is very low to produce any disturbance of the incident electron beam and of the measurement process (Zarbout et al., 2008). Then, the good stability and reliability of the SEM ensure (for a biased voltage applied to the secondary electron collector greater than about 50 V) the current complementarity according to the Kirchhoff law:

$$I_p(t) = I_{ind}(t) + I_\sigma(t) \tag{8}$$

Hence, the presence of the collector with a sufficient biased voltage allows accurate measurements of the currents and therefore the precise determination of the quantity of trapped charges. Incidentally, this makes also possible a precise calculation of the SEE yield σ(t):

$$\sigma(t) = \frac{I_\sigma(t)}{I_p(t)} = \frac{I_p(t) - I_{ind}(t)}{I_{ind}(t) + I_\sigma(t)} = 1 - \frac{I_{ind}(t)}{I_{ind}(t) + I_\sigma(t)} \tag{9}$$

The SEE yield could be deduced directly from the secondary electron current $I_\sigma(t)$ and the primary one $I_p(t)$. However, the use of $I_\sigma(t)$ and $I_{ind}(t)$ in Eq. 9 is more appropriate as it provides the opportunity to circumvent any uncontrolled fluctuation of the primary electron current during the different phases of the experimental process.

It is worth noting that after the fabrication process of polycrystalline samples (with firing temperatures above 1863 K) or the thermal treatment of single crystals (at 1773 K for 4 hours), the final thermal cleaning stage under vacuum of all the samples prior to electron irradiation will completely remove initial charges and surface contamination that could interfere with the generated charges.

4.3 The charging kinetic

The measurement of the foregoing currents gives means to follow the net quantity of trapped charges during irradiation. The current curves $I_\sigma(t)$ and $I_{ind}(t)$ of Fig 5 are a typical example of the recorded currents. As pointed out in the experimental conditions, irradiation is performed with a 10 keV electron beam energy, which ensure a net amount of positive charges in the sample.

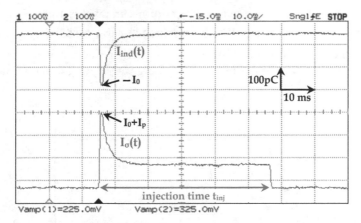

Fig. 5. Current curves recorded during the first injection in polycrystalline alumina sample (d = 4.5 μm) at 473 K. The currents, $-I_0$ = − 225 pA and $I_0 + I_p$ = 325 pA, are measured after a short lag time. The irradiation conditions are: E_p = 10 keV, I_p = 100 pA, t_{inj} = 5 ms, Q_{inj} = 5 pC and J = 4×10^4 pA/cm².

At the beginning of injection, the current curves are affected by the rise time of the amplifiers. Therefore, a short lag time is required to display the actual current values $-I_0$ and $I_0 + I_p$. As irradiation proceeds, the induced current, I_{ind}, increases from $-I_0$ to zero. Concurrently and since the current complementarity conditions are verified (as revealed by Eq. 8), the total secondary electron current I_0 decreases from ($I_0 + I_p$) to I_p. This decrease has been associated with a progressive accumulation of positive charges that are distributed over a depth of about the escape length of secondary electrons λ. As suggested, this could be assigned to recombination of electrons with holes (in this zone), which should be otherwise emitted (Cazaux, 1986) or to field effect (Blaise et al., 2009). The other important feature is indeed that during injection, a negative charge distribution develops in the vicinity of the penetration depth of primary electrons R_p. As result, an internal electric field is established between the two charge distributions (holes near the surface and electrons around R_p) whose direction is oriented towards the bulk. Since the diameter Φ of the irradiated area is much larger than the penetration depth, a planar geometry can be used to evaluate this field (Aoufi & Damamme, 2008) from the Gauss theorem:

$$E(z,t) = \frac{-1}{\varepsilon_0 \varepsilon_r (1 + \varepsilon_r)} \left(1 + \frac{\Phi^2}{4(1 + \varepsilon_r)e^2}\right)^{-1} \int_0^e \rho(z,t)dz + \frac{1}{\varepsilon_0 \varepsilon_r} \int_0^z \rho(z,t)dz \qquad (10)$$

In this expression, where the contribution of image charges has been taken into account, e is the sample thickness, ε_0 is the vacuum permittivity, ε_r the relative permittivity (taken equal to 10 for α-alumina) and $\rho(z, t)$ is the density of charges at the depth z. The first term corresponds to the surface electric field E (0, t).

When I_{ind} reaches a zero value ($I_0(t) = I_p(t)$), a steady state, which corresponds in fact to a self regulated flow regime, is achieved. There, on the average, for each incident electron one secondary electron is emitted. This state is characterized by some constant value of the electric field as well as by a maximum amount of net trapped charge in the sample, Q_{st}, equal to $-\int_{0}^{t_{inj}} I_{ind}(t)\,dt$. The steady state is interesting because it will be taken as a reference state and then the practical choice of the injection time t_{inj} is determined by the achievement of this state. The evolution of the net trapped charges Q_t during injection, which can be deduced from Eq. 7, is shown in Fig. 6.

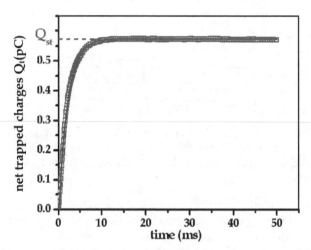

Fig. 6. Evolution with time, at 473 K, of the net charge, Q_t, during the first pulse injection in polycrystalline alumina sample (d = 4.5 μm). The quantity Q_t is derived from the currents of Fig. 5 via Eq. 7. The solid line represents the exponential fit of the data, as given by Eq. 11.

The best fit of data represented in Fig. 6, leads to an exponential time evolution:

$$Q_t(t) = Q_{st}\left[1 - \exp\left(-\frac{t}{\tau_c}\right)\right] \tag{11}$$

where τ_c is the charging time constant (found equal to 2.07×10^{-3} s), which characterizes the charging kinetic of the material in the used experimental conditions.

5. Measurement of the ability to spread charges

The method described above gives the opportunity to evaluate the quantity of trapped charges during irradiation. If the probed zone is initially uncharged, $Q_t(t)$ characterizes the quantity of charges that accumulates after an irradiation time t. Depending on the insulator conduction properties, the accumulated charges can either remain localized or partially (or even totally) spread out of the irradiated volume (discharge phenomenon). Hence, the measurement of the ability of the material to spread charges is of technological interest. In this paragraph, we give details of the experimental procedure developed for the measurement of this ability and we define a recovery parameter allowing the quantitative evaluation of the extent of discharge.

5.1 Experimental protocol

To characterize the degree of spreading (of discharge), we set up a protocol allowing its evaluation. The procedure consists in analysing the states of charging deduced from two pulse electron injections over the same area, separated by some lapse of time as explained in the three following stages.

5.1.1 The first charging stage (first pulse injection)

The first stage is intended to achieve a charging reference state, which is attained once σ reaches the constant value of 1, i.e. the situation where for each electron entering one is emitted. This charging reference state is associated with the trapping of a maximum quantity of charges Q_{st} (cf. Fig. 6). The embedded charges can be either indefinitely trapped or just localised for some time lapse (seconds, minutes, hours or more). If all the charges were trapped in a stable way, the foregoing reference state would stay unchanged when irradiation is turned off. In the case of partially localised charges, some fraction of Q_{st} could manage to spread out from irradiated region, through conduction. To take into account this fact, a pause time long enough to appraise the degree of charge spreading is chosen for the pause stage.

5.1.2 The pause stage

During the pause time Δt, the stability of the amount of charges Q_{st} is determined by the efficiency of traps associated with impurities and lattice defects as well as the charge transport properties. Then, if one wants to evaluate the extent of discharging, one has to perform, with identical conditions, a second injection over the same area as the first one for the purpose of restoring the reference steady state.

5.1.3 The third stage (second pulse injection)

At the inception of the second injection, if the recorded currents are in the continuity of those of the first injection, all the charge Q_{st} have remained trapped in the irradiated volume during the pause time Δt. If it is not the case, a fraction of the charge Q_{st} has been removed from the irradiated zone as a result of discharging (i.e., detrapping and transport) during the pause time.

For the purpose of illustration, the curves of Fig. 7 display the recorded currents during the second injection, performed over the same area as the first injection and with identical experimental conditions, after a pause time $\Delta t = 900$ s. Here, the current curves are not in the continuity of those obtained at the end of the first injection (Fig. 5). As the second injection proceeds, the reference steady state is reached again. The ensuing quantity of net charge introduced during this stage is interpreted as the amount Q_d that has been removed, given by:

$$Q_d = \left[-\int_0^{t_{inj}} I_{ind}(t)\, dt \right]_{second\ injection} \tag{12}$$

In other words, this interpretation means that any loss of charges, during the pause, can be compensated by those introduced during the second irradiation to restore Q_{st}.

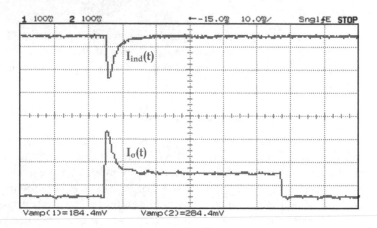

Fig. 7. Current curves obtained during the second pulse injection performed after a pause time Δt = 900 s. Irradiation is carried out over the same area as in the first injection under identical experimental conditions (given in Fig. 5).

The three stages are summarized in Fig. 8, which illustrates the evolution of the amounts of charges remaining in the irradiated volume in the polycrystalline sample of grain diameter d = 4.5 μm at T = 473 K. The data concern the example of Figs. 5 and 7, for which some discharging occurs during the pause Δt. During the pause time t_p ($0 \leq t_p \leq$ Δt), the net charge that still remains in the irradiated volume $Q_l(t_p)$ evolves from its initial value, $Q_l(t_p = 0)$ = Q_{st}, to the final one, $Q_l(t_p =$ Δt$)$ = Q_f. The form of the curve representing $Q_l(t_p)$ will be justified below.

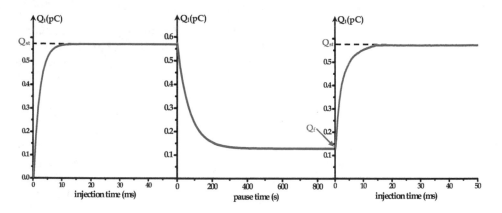

Fig. 8. Illustration of the time evolution of the amounts of charges remaining in the irradiated volume during the three stages. The first and the third curves correspond to Figs. 5 and 7 respectively. The net charge $Q_l(t_p)$ during the pause evolves from Q_{st} to Q_f (which remains in the irradiated volume after a pause of Δt =900 s) as justified below by Eq. (15).

5.2 Definition of a recovery parameter for the evaluation of discharge

We first try to find a kinetic description of the charges that remain in the irradiated volume after a pause time t_p. This amount is given by:

$$Q_1(t_p) = Q_{st} - Q_d(t_p) \qquad (13)$$

where, $Q_d(t_p)$ is given by Eq. 12.

As an example, in Fig. 9, we report the best fit of the evolution of Q_d with t_p, for data obtained, at 473 K in polycrystalline alumina sample (d = 4.5 µm), by performing the measurements over different zones that are sufficiently distant from one another to avoid any overlaps of irradiated volumes.

The fit is well described by an exponential with a time constant τ_r:

$$Q_d(t_p, T) = Q_\infty(T)\left[1 - \exp\left(-\frac{t_p}{\tau_r(T)}\right)\right] \qquad (14)$$

In this equation, Q_∞ is the asymptotic value reached by Q_d. In fact, it stands for the maximum amount of charges going out of the irradiated volume in the discharging process at temperature T. In the example of Fig. 9 where T = 473 K, Q_∞ is found equal 0.46 pC corresponding to about 80 % of Q_{st} (hence, 20 % of the net charge is still inside the irradiated volume). This is an indication of the existence of different trapping sites and that the temperature of 473 K is too low to provoke detrapping from the deeper ones.

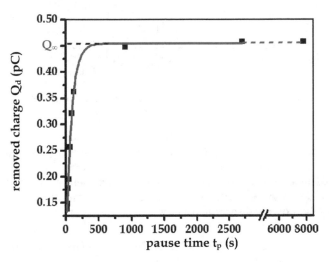

Fig. 9. Evolution with the pause time at T = 473 K of the amount of charges that is removed from the irradiated volume Q_d (in polycrystalline sample of 4.5 µm grain size). The solid line is the exponential fit of the data (Eq. 14). The asymptotic value $Q_\infty(T)$ is attained after a pause time of only 300 s.

From Eq. 14, we can associate to $Q_\infty(T)$ the amount of charges $Q_f(T)$ that still remain in the irradiated volume, $Q_\infty(T) = Q_{st} - Q_f(T)$. Therefore, the remaining quantity of charges $Q_l(t_p, T)$ can be obtained from Eqs. 13 and 14:

$$Q_l(t_p, T) = (Q_{st} - Q_f(T)) \exp\left(-\frac{t_p}{\tau_r(T)}\right) + Q_f(T) \tag{15}$$

The first term in this expression, which corresponds to the curve of the pause stage in Fig. 8, expresses charge decay under the internal electric field. Generally, the time constant τ_r can be set equal to ε/γ, where ε is the dielectric permittivity ($\varepsilon = \varepsilon_r\varepsilon_0$) and γ is the electric conductivity of the material (Adamiak, 2003; Cazaux, 2004). The value of τ_r deduced from Fig. 9 is 82 s giving a conductivity of about 10^{-14} $\Omega^{-1}\text{cm}^{-1}$ which can be expected for this material, in agreement with the experimental value of the resistivity obtained for this sample in our laboratory (1.2 10^{14} Ωcm).

The asymptotic value Q_∞ at T = 473 K is reached after only 300 s (Fig. 9). Therefore, one can anticipate that at temperatures within the range of interest (300 – 663 K), the condition for reaching the asymptotic value $Q_\infty(T)$ are met for the chosen pause time Δt of 900 s.

The measured value of $Q_\infty(T)$ is the result of detrapping of charges and their subsequent transport under the internal electric field. During this transport, a fraction of the detrapped charges can undergo a retrapping in deeper traps in the irradiated volume and eventually a recombination. The overall effect is a variation of charge distribution in the volume of interest, which affects the electric field. Consequently, since in the considered temperature range the experimental results do not reveal any significant dependence of Q_{st} on temperature, the ratio $R(T) = Q_\infty(T)/Q_{st}$ can be expressed in terms of the measured currents:

$$R(T) = \frac{Q_\infty(T)}{Q_{st}} = \frac{-\left[\int_0^{t_{inj}} I_{ind}(t)dt\right]_{second\ injection}}{-\left[\int_0^{t_{inj}} I_{ind}(t)dt\right]_{first\ injection}} \tag{16}$$

This experimental parameter, which measures the fraction of charge removed from the irradiated volume, also characterizes the extent of discharging.

The ratio $R(T)$, which can vary between 0 and 1, corresponds to an evolution from either the dominance of stable charge trapping (low values of R) or of charge spreading (high values of R, with R = 1 for a complete recovery of the uncharged state). The rate at which charges are detrapped depends usually on an attempt escape frequency and an activation energy linked to the trap depth. As a result, one can expect that, the variation of R with temperature could shed some light on the discharge process.

6. Effect of microstructure induced by sintering on the ability of a dielectric to trap or spread charges

As it will be pointed out, trapping and spreading are intimately linked to the microstructure and defects. Sintering not only leads to the creation of new interfaces but also to important phenomena such as segregation at grain boundaries and defects association.

6.1 Charge spreading in reference materials (single crystals)

The results of the two types of single crystals are reported in Fig. 10. The values of R in both Pi-Kem and RSA are zero up to 473 K, indicating a perfect stable trapping behavior. Above 473 K, R increases but the enhancement is significant only for Pi-Kem.

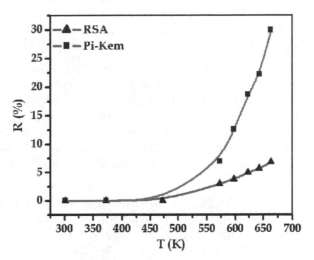

Fig. 10. Fraction R of charges removed from the irradiated volume as a function of temperature for the two types of single crystals (Pi-Kem and RSA samples).

In order to interpret these behaviors we can consider the following results:

i.   Cathodoluminescence spectra obtained in similar Pi-Kem samples (Jardin et al., 1995) have identified mainly the F and F+ centers. Stable trapping in these centers is expected due to their deep level in the band gap (3 eV, and 3.8 eV for the F and F+, respectively). The increase of R above 473 K might be due to the intervention of excited F centers whose energy levels are believed very near the edge of the conduction band (Bonnelle & Jonnard, 2010; Jonnard et al. 2000; Kröger, 1984).

ii. The RSA material displays a more stable trapping ability than the Pi-Kem one above 473 K. This stable trapping raises queries about the role of the defects induced by the impurities (Table 1) and in particular those by the dominant silicon. With such amount of silicon (290 ppm), the concentration of the defects associated with Si exceeds the others. Consequently, one can deduce that stable trapping may occur on these defects.

As mentioned in paragraph 2, the dissolution of silicon into Al_2O_3, is expected to be compensated by a negatively charged cationic vacancy, V_{Al}''' (Eq. 4). In this context, the positively charged substitutional silicon $Si_{Al}^{\bullet}$ may act as electron trapping site while the cationic vacancy V_{Al}''' as hole trap. Upon trapping one electron during irradiation, $Si_{Al}^{\bullet}$ induces a donor level associated to Si_{Al}^{X}, which is estimated at 1.59 eV below the edge of the conduction band (Kröger, 1984). With regard to V_{Al}''', hole trapping will give an acceptor level (associated to V_{Al}'') located at 1.5 eV above the valence band (Kröger, 1984).

Accordingly, with such relatively deep trap levels, it is very unlikely that detrapping of charge carriers occurs at the temperatures of our experiments. However, the contribution of the other impurities, such as the ones of smaller valence than the host, cannot be ruled out because trapping depends not only on the defect concentration but also on their specific trapping properties such as the capture cross section of traps which is mainly determined by their charge state. Indeed, the cathodoluminescence spectra have also detected (Jardin et al., 1995) in similar RSA samples (which like ours were annealed at 1773 K during 4 hours) the deep centers F_{cation}, such as F^x_{Mg} with an energy level at 4 eV below the edge of the conduction band (Kröger, 1984).

We alternatively tried to shed some more light on the trapping behaviors of single crystals by using the Scanning Electron Microscope Mirror Effect "SEMME" method, which requires net negative charging (Liebault et al., 2003; Vallayer et al. 1999). Thus, we performed electron injection with 30 keV (energy greater than E_{pII}) focused beam and an injected dose of 300 pC followed by a scanning of the sample surface, with low electron beam energy (some hundred of eV). In RSA samples we find a mirror image at 300 K, which remains stable up to about 663 K. In contrast, in Pi-Kem materials no mirror image formation was achievable even at 300 K.

It must be reminded that in the SEMME method the observation of a mirror image, immediately after irradiation, is due to the presence of a sufficiently high concentration of stable traps for electrons. Thus, no mirror image is observed in Pi-Kem samples as they do not contain enough traps. The fact that in RSA the mirror is maintained at high temperatures confirms the presence of deep traps and somewhat supports the assumption, discussed above in point (ii), that stable trapping can be assigned to the defects induced by the dissolution of the dominant silicon impurity.

In contrast with SEMME method, the ICM method can give, during charging, an induced current whatever the trap concentration is. So, the parameter R gives indications on the stability of trapped charges, after a pause Δt, independently of the trap concentration. As a result, at temperatures lower than 573 K, for which trapped charges are stable in RSA and Pi-Kem samples, the R values are very low (R = 0 at room temperature for both kinds of single crystals). Above 573 K, RSA samples have R values lower than those of Pi-Kem. This is a supplementary indication of the existence of deeper traps in RSA, which is one of the requirements for the mirror image formation at high temperatures in the SEMME method.

6.2 Charge spreading in sintered alumina of various microstructures and impurities

6.2.1 Charge spreading in pure sintered alumina

The values of R(T) for the three polycrystalline alumina pure samples as function of temperature are shown in Fig. 11. The comparison with single crystals (R close to zero up to 573 K) indicates that the presence of grain boundaries makes trapping less stable in polycrystalline alumina (R is substantially above zero). This interpretation agrees with the fact that, with SEMME method, we do not detect any mirror image in the polycrystalline samples.

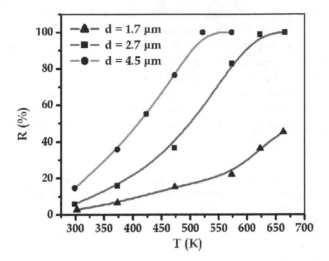

Fig. 11. Fraction R of charges removed from the irradiated volume as function of temperature for pure polycrystalline alumina samples of different grain diameters d.

In the RSA single crystal and pure polycrystalline alumina, silicon is the dominant impurity. Therefore, one can suggest that atomic disorder introduced by grain boundaries gives rise to states closer to the edge of the conduction band than those of Si_{Al}^{x} and $V_{Al}^{'''}$, which are most likely responsible for the stable trapping in the bulk.

For fine grains (d = 1.7 μm), R spans from 0 (at 300 K) to 45 % (at 663 K) whereas for large grains (d = 4.5 μm) R varies between 15 % (at 300 K) to near complete discharge (about 100 %) above 473 K. One would expect the behaviour of polycrystalline material to tend towards that of single crystals when the grain size increases. However, the reverse is clearly seen. The explanation of this apparent contradiction can be found by considering the distribution of impurities in the polycrystalline samples. We have to bear in mind that the microstructural evolution during the sintering process, with in particular the achievement of a given grain size, is concomitant with segregation to grain boundaries of impurities, which corresponds to a purification of the grain (interpreted as an internal gettering). In fact, the large grain size of 4.5 μm has required both a higher sintering temperature and a longer dwelling time than those used to attain 1.7 μm. Hence, the enhancement of gettering effect in large grain sample may have lowered the concentration of deep traps (such as Si_{Al}^{x} and $V_{Al}^{'''}$) in the bulk. In this assumption, the impurities (mainly Si) segregates at grain boundaries where they can be associated to other defects to form clusters, as suggested by positron measurements (cf. paragraph 3), which may be less efficient in trapping charges.

This agrees with the experimental facts:

i.   The R values are always higher for the larger grain (for example at 573 K, R = 100 % for d = 4.5 μm and only 22 % for d = 1.7 μm). This means that the density of deep traps in large grains is substantially lower.

ii.   The foregoing assumption is somewhat confirmed in Fig. 12, where below 573 K the semi-logarithmic plot of R versus reciprocal temperature exhibits an Arrhenius law leading to an activation energy about 0.12 eV whatever the grain size is. This same activation energy means that we are dealing with detrapping from similar trapping sites (i.e., grain boundary traps). The continuous variation in R over a large temperature range, as shown in Fig 11, indicates a detrapping from a density of continuous trapping states rather than from a single trapping level. This aspect characterizes disordered solids in which hopping conduction mechanism can occur with the same order of magnitude of activation energy (Blaise, 2001).

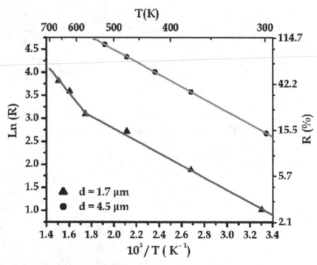

Fig. 12. Semi-logarithmic plot of the ratio R expressing the degree of discharge versus reciprocal temperature for polycrystalline alumina (solid line: linear fit of the data). For T below 573 K, discharging is characterized by the same activation energy (0.12 eV). Above 573 K, a second energy (0.26 eV) in small grains sample arises.

iii. for temperature above 573 K the smallest grain size sample presents a second detrapping zone corresponding to an activation energy of about 0.26 eV (i.e., twice the energy at lower temperature). This fact is a further confirmation of the presence in the smaller grain size of a higher density of deeper traps located within the grain (likely $Si_{Al}^{\times}$ and V_{Al}''') in accordance with a less efficient gettering effect.

6.2.2 Charge spreading in impure sintered alumina

The results of the three polycrystalline alumina impure samples are reported in Fig. 13. The comparison of the pure and impure polycrystalline samples reveals a more stable trapping behavior in impure samples in the whole temperature range. Furthermore, two opposite behaviors arise: stable trapping increases with the grain diameter in impure samples (Fig.13) and the contrary is obtained in the pure ones (Fig. 11). In the impure material, the contents of impurities are much higher and expected above the solubility limits. Hence, gettering might be less efficient due to a possible saturation of grain boundaries. In addition, interactions between the various defects generated by the foreign elements are expected (Gavrilov et al., 1999). It must be pointed out that the great variety of impurities and their substantial content make difficult the interpretation of the results due to a possible interference of co-segregation, which is difficult to predict when more than three elements are involved. The actual situation is even more complicated by the fact that segregation leads to the creation of a space charge at grain boundaries (Tiku & Kröger, 1980) with a sign that depends on the segregated impurities, which may interfere with the charging process.

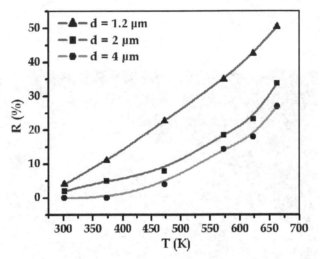

Fig. 13. Fraction R of charges removed from the irradiated volume as a function of temperature for impure polycrystalline alumina samples of different grain diameters d.

The semi-logarithmic plot of the degree of discharge R versus reciprocal temperature exhibits, as for the pure samples, an Arrhenius type law leading to an activation energy of about 0.12 eV for the smallest grain size sample (d = 1.2 µm) and about 0.28 eV for the largest one (d = 4 µm). These activation energies are an indication of detrapping from trapping sites located within the grain boundary.

It is worth noting that, at room temperature and in the same impure polycrystalline samples, breakdown strength increases with the grain sizes (Liebault, 1999; Si Ahmed et al., 2005), which is also the case for the fraction of removed charges R. Therefore this correlation confirms the importance of charge spreading to prevent breakdown.

6.3 Charge spreading in alumina of sub-micrometric grain size

The evolution of the properties with the grain size raises queries about the effect of changing the grain size from micron to nanometer scales. The study of charging properties of nanostructured alumina is beyond the scope of the present chapter. However, there is interest in trying to verify whether they can be obtained by simple extrapolation from the results of micrometric grain size materials.

Nanopowders, with grain diameter of about 27 nm, have been synthesized by the gaseous phase method and compacted via magnetic compaction process at Ural State Technical University, Russia (Kortov et al., 2008). Next, sintering of the compacts has been carried out at Institut National des Sciences Appliquées (INSA) of Lyon (France). The sintering temperature of 1473 K (dwelling time 60 min) was reached at a rate of 3 K/min. As expected, with such heating rate, a substantial grain growth occurred (i.e., from 27 to about 100 nm, cf. Fig. 14). Indeed, grain growth could have been reduced by using faster heating rates that are made possible by different sintering techniques such as spark plasma sintering.

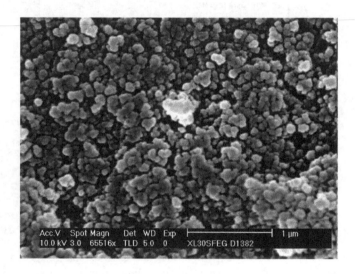

Fig. 14. Microstructure of the sub-micrometric grain size of the impure polycrystalline alumina sample after sintering. This picture is the SEM observation of fracture surface. The average grain diameter has grown after sintering to about of 100 nm (the initial particle diameter prior to sintering was about 27 nm).

The overall purity of this material is about the same as the one of impure polycrystalline alumina. In Fig. 15, the fraction R of charges removed from the irradiated volume as a function of temperature is given for the 0.1 and 1.2 μm samples. The manifest difference is the sharp enhancement of R between 600 and 663 K for the 0.1 μm sample (the value of R increases from 30 to 90 %), which contrasts with the continuous behaviour of the other polycrystalline samples. The activation energy that arises from the semi-logarithmic plot of the recovery parameter versus reciprocal temperature in the range 600-663 K is 0.53 eV (Moya et al., 2007). These results are an indication that detrapping occurs, at about 600 K, from a dominant efficient trap having a well-defined energy level in the gap as in the case of silver doped single crystal (Zarbout et al. 2010).

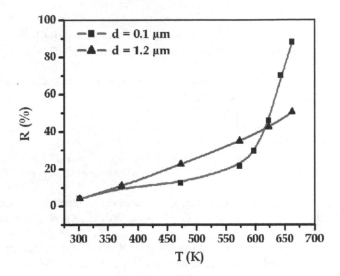

Fig. 15. Fraction R of charges removed from the irradiated volume as a function of temperature for the sub-micrometric and 1.2 μm grain sizes alumina samples.

7. Conclusion

This chapter provides a method for the characterization of charge trapping and spreading in dielectrics. A quantitative recovery parameters reflecting the relative degree of the two competing processes is accurately derived. The experimental set up makes possible the assessment of the effect of temperature (in the range 300-700 K). The ability of polycrystalline alumina to trap or, conversely, to spread charges depends strongly on the grain size and segregation of impurities at interfaces. The results suggest that the grain boundary interfaces can be associated to shallow traps whereas the defects within the grains to deeper ones. The strong tendency for segregation of the main impurities implies that an internal gettering effect can also intervene. It appears therefore that the control of the microstructural development, during the conventional sintering process, is of importance as it provides ways to influence the insulator properties in technological applications of oxide ceramics, for instance, the breakdown strength. Further investigations dealing with the properties of nanostructured materials, processed by sintering techniques that reduce grain growth, could bring more understanding of the role of interfaces.

8. Acknowledgments

The authors are grateful to Dr. Goeuriot D. and Dr. Liébault J. (E.N.S.M.), to Prof. Kortov V.S. (Ural State Technical University) for the supply of some samples. Fruitful discussions with Dr. Bernardini J. (Im2np) and Prof. Fakhfakh Z. (LaMaCoP) were highly appreciated. The first author gratefully acknowledges financial support from the Ministry of Higher Education and Scientific Research of Tunisia and the French Institute of Cooperation.

9. References

Adamiak, K. (2003). Analysis of charge transport in high resistivity conductors under different conduction models. *Journal of Electrostatics*, Vol. 57, No. 3-4, (March 2003), pp. 325-335, ISSN 0304-3886

Aoufi, A. & Damamme, G. (2008). Analysis and numerical simulation of secondary electron emission of an insulator submitted to an electron beam. *Proceedings of ISDEIV 2008 23th International Symposium on Discharges and Electrical Insulation in Vacuum*, pp. 21-24, ISBN 978-973-755-382-9, Bucharest, Romania, September 15–19, 2008

Atkinson, K. J. W.; Grimes, R. W.; Levy, M. R.; Coull, Z. L. & English, T. (2003). Accommodation of impurities in α-Al$_2$O$_3$, α-Cr$_2$O$_3$ and α-Fe$_2$O$_3$. *Journal of the European Ceramic Society*, Vol. 23, No. 16, (December 2003), pp. 3059-3070, ISSN 0955-2219

Blaise, G. & Le Gressus, C. (1991). Charging and flashover induced by surface polarization relaxation process. *Journal of Applied Physics*, Vol. 69, No. 9, (May 1991), pp. 6334-6339, ISSN 0021-8979

Blaise, G. (2001). Charge localization and transport in disordered dielectric materials. *Journal of Electrostatics*, Vol. 50, No. 2, (January 2001), pp. 69-89, ISSN 0304-3886

Blaise, G.; Pesty, F.; & Garoche, P. (2009). The secondary electron emission yield of muscovite mica: Charging kinetics and current density effects. *Journal of Applied Physics*, Vol. 105, No. 3, (February 2009), pp. 034101(1-12), ISSN 0021-8979

Bonnelle C. & Jonnard P. (2010). Dynamics of charge trapping by electron-irradiated alumina. *Physical Review B.*, Vol. 82, No. 7, (August 2010), pp. 075132(1-8), ISSN 1098-0121

Cazaux, J. (1986). Some considerations on the electric field induced in insulators by electron bombardment. *Journal of Applied Physics*, Vol. 59, No. 5, (March 1986), pp. 1413-1430, ISSN 0021-8979

Cazaux, J. (2004). Charging in Scanning Electron Microscopy "from Inside and Outside". *Scanning*, Vol. 26, No. 4, (July/August 2004), pp. 181-203, ISSN 1932-8745

Chiang, Y. -M.; Birnie III, D. & Kingery, W. D. (1996). *Physical Ceramics: Principles for Ceramic Science and Engineering*, (Wiley), John Wiley & Sons, ISBN 0-471-59873-9, New York

Dörre, E. & Hübner, H. (1984). *Alumina: processing, properties, and applications*, Springer-Verlag, ISBN 3-540-13576-6, Berlin

Gavrilov, K. L.; Bennison, S. J.; Mikeska, K. R. & Levy-Setti, R. (1999). Grain boundary chemistry of alumina by high-resolution imaging SIMS. *Acta Materialia*, Vol. 47, No. 15-16, (November 1999), pp. 4031-4039, ISSN 1359-6454

Gontier-Moya, E. G.; Bernardini, J. & Moya, F. (2001). Silver and Platinum diffusion in alumina single crystals. *Acta Materialia*, Vol. 49, No. 4, (autumn 2001), pp. 637-644, ISSN 1359-6454

Haddour, L.; Mesrati, N.; Goeuriot, D. & Tréheux, D. (2009). Relationships between microstructure, mechanical and dielectric properties of different alumina materials. *Journal of the European Ceramic Society*, Vol. 29, No. 13, (October 2009), pp. 2747-2756, ISSN 0955-2219

Jardin, C.; Durupt, P.; Bigarre, J. & Le Gressus, C. (1995). The surface potential and defects of insulating materials probed by electron and photon emissions. *Annual Report Conference on Electrical Insulation and Dielectric Phenomena*, pp. 548-551, ISBN 0-7803-2931-7, Virgina Beach, VA, USA, October 22-25, 1995

Jonnard, P.; Bonnelle, C.; Blaise, G.; Remond, G. & Roques-Carmes, C. (2000). F+ and F centers in α-Al₂O₃ by electron-induced x-ray emission spectroscopy and cathodoluminescence. *Journal of Applied Physics*, Vol. 88, No. 11, (December 2000), pp. 6413-6417, ISSN 0021-8979

Kansy, J. (1996). Microcomputer program for analysis of positron annihilation lifetime Spectra. *Nuclear Instruments and Methods in Physics Research Section A: Accelerators, Spectrometers, Detectors and Associated Equipment*, Vol. 374, No. 2, (May 1996), pp. 235-244, ISSN 0168-9002

Kolk, J. & Heasell, E. L. (1980). A study of charge trapping in the Al-Al₂O₃-Si, MIS system. *Solid State Electron*, Vol. 23, No. 2, (February 1980), pp. 101-107, ISSN 0038-1101

Kortov, V. S.; Ermakov, A. E.; Zatsepin, A. F. & Nikiforov S. V. (2008). Luminescence properties of nanostructured alumina ceramic. *Radiation Measurements*, Vol. 43, No. 2-6, (February-June 2008), pp. 341 – 344, ISSN 1350-4487

Krause-Rehberg, R. & Leipner, H. S. (1999). *Positron Annihilation in Semi-conductors: defect studies*, Springer-Verlag, ISBN 3-540-64371-0, Berlin

Kröger, F. A. (1984). Electrical properties of α-Al₂O₃, In: *Advances in Ceramics, Vol. 10: Structure and Properties of MgO and Al₂O₃ Ceramics*, W. D. Kingery, pp. 1-15, American Ceramic Society, ISBN 0916094626, Columbus, OH

Lagerlöf, K. P. D. & Grimes, R. W. (1998). The defect chemistry of sapphire (α-Al₂O₃). *Acta Materialia*, Vol. 46, No. 16, (October 1998), pp. 5689-5700, ISSN 1359-6454.

Le Gressus, C.; Valin, F.; Henriot, M.; Gautier, M.; Duraud, J. P.; Sudarshan, T. S.; Bommakanti, R. G. & Blaise G. (1991). Flashover in wide-band-gap high-purity insulators: Methodology and mechanisms. *Journal of Applied Physics*, Vol. 69, No. 9, (May 1991), pp. 6325-6333, ISSN 0021-8979

Liebault, J. (1999). Ph. D. thesis. *Behavior of alumina materials during injection of charges. Relation between microstructure, dielectric breakdown and image current measurement (The SEMM method)*, Ecole Nationale Supérieure des Mines de Saint Etienne, France, INIST T 126061

Liebault, J.; Vallayer, J.; Goeuriot, D.; Tréheux, D. & Thévenot, F. (2001). How the trapping of charges can explain the dielectric breakdown performance of alumina ceramics. *Journal of the European Ceramic Society*, Vol. 21, No. 3, (March 2010), pp. 389-397, ISSN 0955-2219

Liebault, J.; Zarbout, K.; Moya, G. & Kallel A. (2003). Advanced measurement techniques of space-charge induced by an electron beam irradiation in thin dielectric layers. *Journal of Non-Crystalline Solids*, Vol. 322, No. 1-3, (July 2003), pp. 213-218, ISSN 0022-3093

Mohapatra, S. K. & Kröger, F. A., (1977). Defect structure of α-Al₂O₃ doped with titanium. *Journal of the American Ceramic Society*, Vol. 60, No. 9-10, (September 1977), pp. 381-387, ISSN 1551-2916

Moya, G. & Blaise, G. (1998). Charge trapping-detrapping induced thermodynamic relaxation processes, In: *Space Charge in Solids Dielectrics*, J.C. Fothergill & L.A. Dissado, pp. 19-28, Dielectric Society, ISBN 0 9533538 0 X, Leicester, UK

Moya, G.; Kansy, J.; Si Ahmed, A.; Liebault, J.; Moya F. & Gœuriot, D. (2003). Positron lifetime measurements in sintered alumina. *Physica Status Solidi (a)*, Vol. 198, No. 1, (July 2003), pp. 215-223, ISSN 1862-6319

Moya, G.; Zarbout, K.; Si Ahmed, A.; Bernardini, J.; Damamme, G. & Kortov, V. (2007). Grain size effect on electron transport properties of poly and nano-crystalline alumina, *First International Meeting on Nano-materials*, Belfast, January 13-15, 2007

Rasmussen, J. J. & Kingery, W. D., (1970). Effect of dopants on the defect structure of single-crystal aluminium oxide. *Journal of the American Ceramic Society*, Vol. 53, No. 8, (August 1970), pp. 436-440, ISSN 1551-2916

Si Ahmed, A.; Kansy, J.; Zarbout, K.; Moya, G.; Liebault, J. & Goeuriot, D. (2005). Microstructural origin of the dielectric breakdown strength in alumina: A study by positron lifetime spectroscopy. *Journal of the European Ceramic Society*, Vol. 25, No. 12, (2005), pp. 2813-2816, ISSN 0955-2219

Song, Z. G.; Ong, C. K. & Gong, H. (1996). A time-resolved current method for the investigation of charging ability of insulators under electron beam irradiation. *Journal of Applied Physics*, Vol. 79, No. 9, (May 1996), pp. 7123-7128, ISSN 0021-8979

Stoneham, A. M. (1997). Electronic and defect processes in oxides. The polaron in action. *IEEE Transactions on Dielectrics and Electrical Insulation*, Vol. 4, No. 5 (October 1997), pp. 604-613, ISSN 1070-9878

Suharyanto, Yamano, Y.; Kobayashi, S.; Michizono, S. & Saito, Y. (2006). Secondary electron emission and surface charging evaluation of alumina ceramics and sapphire. *IEEE Transactions on Dielectrics and Electrical Insulation*, Vol. 13, No. 1 (February 2006), pp. 72-78, ISSN 1070-9878

Thome, T.; Braga, D. & Blaise, G. (2004). Effect of current density on electron beam induced charging in sapphire and yttria-stabilized zirconia. *Journal of Applied Physics*, Vol. 95, No. 5, (March 2004), pp. 2619-2624, ISSN 0021-8979

Tiku, S. K. & Kröger, F. A. (1980). Effects of space charge, grain-boundary segregation, and mobility differences between grain boundary and bulk on the conductivity of polycrystalline Al_2O_3. *Journal of the American Ceramic Society*, Vol. 63, No. 3-4, (March 1980), pp. 183-189, ISSN 1551-2916

Touzin, M.; Goeuriot, D.; Guerret-Piécourt, C.; Juvé, D. & Fitting, H. -J. (2010). Alumina based ceramics for high-voltage insulation. *Journal of the European Ceramic Society*, Vol. 30, No. 4, (March 2010), pp. 805-817, ISSN 0955-2219

Vallayer, B.; Blaise, G. & Treheux, D. (1999). Space charge measurement in a dielectric material after irradiation with a 30 kV electron beam: Application to single-crystals oxide trapping properties. *Review of Scientific Instruments*, Vol. 70, No. 7, (July 1999), pp. 3102-3112, ISSN 0034-6748

Zarbout, K.; Moya, G. & Kallel, A. (2005). Determination of the electron beam irradiated area by using a new procedure deriving from the electron beam lithography technique. *Nuclear Instruments and Methods in Physics Research Section B: Beam Interactions with Materials and Atoms*, Vol. 234, No. 3, (June 2005), pp. 261-268, ISSN 0168-583X

Zarbout, K.; Si Ahmed, A.; Moya, G.; Bernardini, J.; Goeuriot, D. & Kallel, A. (2008). Stability of trapped charges in sapphires and alumina ceramics: Evaluation by secondary electron emission. *Journal of Applied Physics*, Vol. 103, No. 5, (March 2008), pp. 054107(1-7), ISSN 0021-8979

Zarbout, K.; Moya, G.; Si Ahmed, A.; Damamme, G. & Kallel, A. (2010). Study of discharge after electron irradiation in sapphires and polycrystalline alumina. *Journal of Applied Physics*, Vol. 108, No. 9, (November 2010), pp. 094109(1-8), ISSN 0021-8979

8

Evaluation of Dielectric Properties from the Cakes of Feldspathic Raw Material for Electrical Porcelain Production

V.P. Ilyina

*Establishment the Karelian Centre of Science of the Russian
Academy of Sciences Institute of Geology of the Russian
Academy of Science
Russia*

1. Introduction

Feldspars are anhydrous alumosilicates containing alkaline (Na^+, K^+) and alkaline-earth (Ca^{2+}) cations. The basic types of feldspars used in ceramic production are: potassic feldspar (microcline) $K_2O \bullet Al_2O_3 \bullet 6SiO_2$, sodic feldspar (albite) $Na_2O \bullet Al_2O_3 \bullet 6SiO_2$ and calcium feldspar (anorthite) $CaO \bullet Al_2O_3 \bullet 2SiO_2$. The dielectrical properties of electrical porcelain depend on the mineralogical composition, alkali ratio and total alkali content of feldspathic raw material [1].

To examine new sources of raw materials for electrical porcelain production, we have evaluated nonconventional types feldspathic rocks of Karelian. Feldspathic rocks such as alkaline and nepheline syenites from the Elisenvaara and Yeletozero deposits, aplite-like granite from the Louhi area and volcanics such as Kostomuksha halleflinta and Roza-Lambi quartz porphyry. On a chemical compound they represent Alumosilikate K, Na, Ca, less often Ba. Form isomorphic numbers, including Plagioclase [3]. In world practice, the above rocks are a common source of mineral products which has some advantages over pegmatite.

Nonconventional types of feldspathic raw materials were evaluated on cakes because the dielectrical and other properties of natural feldspars and their cakes are much the same, while the chemical bonds of natural feldspars persist after their melting and are inherited by the glass phase of the resulting ceramics [1,2].

Cakes (material in vitreous state, as in ceramics) were prepared from finely ground (particle size 0.063 mm) powders of deferrized feldspathic rocks by caking them in crucibles at 1350° C for 3 hr.

We have assessed the electrical properties (dielectrical permeability -ε, dielectrical loss angle tangent -tgδ and electrical resistance -lgρ), the thermal coefficient of linear expansion (TCLE) and pH of nonconventional types of feldspathic raw materials such as potassic halleflinta, quartz porphyry, syenite and granite-aplite and compared them with those of pegmatite, a common paw material for ceramics production. The compositions and physico-technical properties of feldspathic rocks are shown in Table 1.

2. Electrical characteristics of feldspathic raw material

To measure electrical properties, we prepared specimens of cakes. They were 20 - 25 mm in diameter and 2 - 3 mm in height. ε-, lgρ- and tg δ-values were measured using a bridge with a capacity E of 7-8 and a working frequency of 1000 Hz at 20°C.

Electrical properties were estimated by introducing an additional coefficient to account for additional capacity on the specimen zones not covered by electrodes using the formulas: dielectric permeability: $\varepsilon = \kappa_1 \kappa_2 c$, where $\kappa_1 = 1.14$ is the coefficient of the sensor, κ_2 is the coefficient of specimen thickness, c is specimen capacity; specific electrical resistance: lgρ = κ/g·l, where κ = 33.55 is the sensor coefficient, g is conductivity, l is specimen thickness; dielectrical loss: tg δ = 0.175·g/s, where c is specimen capacity and g is conductivity.

The results of the measurement of the dielectric properties of feldspathic rocks are shown in Table 1, and the dependence of variations in ε, lgρ and tgδ on the quantities of microcline and plagioclase in the cake is shown in Figures 2.

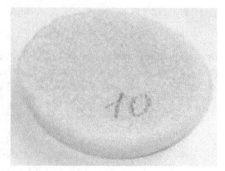

Fig. 1. Cake from feldspathic raw material.

Dielectrical permeability, ε, is the electrical parameter of a mineral showing its ability to polarize in an electrical field. According to experimental data, the dielectrical permeability of the cakes of the rocks evaluated (Nos. 1 – 12) varies from 3.26 to 8.1. The dielectrical permeabilities of feldspathic rocks from other deposits (Nos. 13 – 18), shown in Table 1, range from 5.6 to 7.7, and according to E.V. Rozhkova's results presented in the literature [4,5] /(Golod et al., 1975), they vary from 5.6 – 6.3. This means that the variation range of ε of the cakes of nonconventional feldspathic rocks is within known values.

Analysis of the dielectrical permeability values obtained has shown that ε depends largely on mineralogical composition. Microcline and quartz were shown to have the greatest effect on dielectrical permeability. A high percentage of quartz (44 mass.%), in contrast to that of other rocks, in volcanics from the Roza-Lambi deposit (cake no. 7) contributes to a decline in ε of the cake to 5.4 dtn/cm².

The ε dielectrical permeability of plagioclase rock cakes (Nos. 10 – 12, Table) is lower than that microcline cakes (3.26-4.02), which seems to be due to the presence of large quantities of quartz and Ca^{2+} and Mg^{2+} ions in cakes 11-12 and Ba^{2+} ions in cakeв no. 10, because the electrical conductivity of K^+ and Na^+ ions in dielectrics is higher than that of Ca^{2+}, Mg^{2+} and Ba^{2+} ions [1,4].

Type of breeds	Feldspar deposits		Mineral structure, mass. %				Dielectric properties			TKL R⊙10^-6 1/ grad	pH
	Area	Deposits	№ппп	Q	Mi	Pl	E	$\rho \odot 10^{10}$ Ohm⊙ cm	tg δ	400°C	
Pegmatite	Chaupino-Louhi	Hetolambino, Uracco, Civ-guba	1 2 3	2,5 28.6 24.2	70.0 64.1 61.3	27.5 7.3 14.5	7.70 7.53 7.35	- 0,94 1.00	- 0.148 0.062	7.70 7.93 7.72	9.67 9.08 9.85
	Priladozhye	Jccima Lypikko	4 5	27.2 27.8	49.8 46.6	23.0 25.6	6.20 6.73	0.95 1.00	0.900 0.127	8.10 8.51	9.83 9.82
	Uljlega	Cjryla	6	7.1	58.0	34.9	7.00	2.10	0.027	7.90	8.84
Volca-nic rock	White Sea	Roza-Lambi	7	44.0	44.2	11.8	5.40	0.85	0.160	8.08	7.10
	Kalevala	Kostomuks ha	8	24.2	68.0	7.8	7.50	1.23	0.050	9.15	10.0
Syenite	Louhi	Yeletozero	9	23.1	53,5	31.4	6.65	1.51	0.032	7.83	9.80
	Priladozhye	Elisenvaara	10	2.8	43,0	54.2	3.26	2.00	0.025	8.00	8.25
Gra-nite-aplite	Louhi	Yeletozero	11	29.1	Or-30,8	40.4	3.87	0.85	0.207	7.89	7.98
		Slyudozero	12	22.0	Or-11,1	66.9	4.02	1.32	0.034	7.84	9.76
Pegmatite-granite	Mamsko-Chauiscoe	Mamskoe,B. Northern	13	9.0	71,0	20.0	7.5	-	-	7.78	-
	Enskoe	Rikolatva	14	10.1	69,5	26.4	7.0	-	-	7.52	8.4
	Central Asia	Ljngarskoe	15	15.5	47.7	36.8	5.6	-	-	7.82	-
	Kazakhstan	Karaotse-lgskoye	16	2.5	58,7	38.8	7.7	-	-	7.76	-
	Uralskoe	Malache-vskoye	17	43.6	29,3	27.1	5.6	-	-	7.80	
	Finland	Kemio (FFF)	18	8.0 50	37,0 50,0	55.0 450	-	-	-	-	8.3 83

Table 1. Mineral structure and properties of feldspar breeds.

Figure 2 shows that high microcline and plagioclase concentrations in feldspathic rocks are responsible for a low angle tangent of dielectrical loss in cakes: (no. 8) potassic halleflinta - 0,050, (no. 10) syenite - 0,025 – 0.032, (no. 6) and pegmatite (Cjryla) – 0.027- and increases their electrical resistance (Fig.3) -(1,23, 2,0, 2,10⊙10^{10} Ohm⊙cm), which is consistent with GOST parameters 2484-85 for electrical porcelain: dielectrical permeability not more than 7, the angle tangent of dielectrical loss – 0.030 and electrical resistance 1-2⊙10^{12}. Low dielectrical loss (0.025) and high electrical resistance (2.0⊙10^{10} Ohm⊙cm) are characteristic of Elisenvaara syenites which have a high total percentage of microcline and plagioclase (97 mass.%).

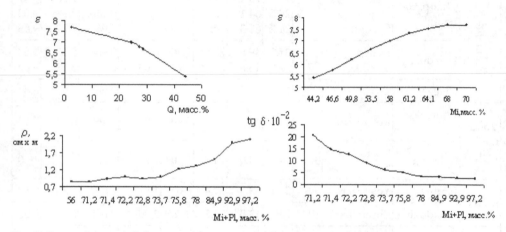

Fig. 2. Dependence of dielectric properties on feldspathic rock composition.

Elisenvaara syenites and Kostomuksha volcanics are most efficient feldspathic raw materials for electrical porcelain production in terms of their electrical properties.

3. Thermal Coefficient of Linear Expansion (TCLR)

Thermal coefficient of linear expansion (TCLR) is a parameter which shows thermal expansion upon burning of ceramic products.

TCLR was measured on a DKV- 4 quartz dilatometer on specimens, 55⊙ × 5 m in size, produced by pressing ground rocks to a grain size of 0.063 mm and subsequent burning in a silitic furnace at 1350°C to reach the maximum density of the specimens. TCLR was estimated using the formula: $\alpha = \frac{1}{l_0} \times \frac{l_t - l_0}{t_1 - t_0} + 55,5 \times 10^{-6} 1 / deg.$, where t_0 is initial measurement temperature, 20°C , t_1 is final measurement temperature, 400, 700°C; l_0 is the original length of the specimen, mm, l_t is specimen lengthening, mm; 5.5⊙(10^{-6}- 1/deg. of the TCLR of quartz glass.

The highest TCLR values (9.15 and 11.75⊙10^{-6} 1/deg., respectively) are characteristic of halleflinta (cake no. 8) at 400° and 700°C, respectively. Halleflinta differs structurally from pegmatite in having finely dispersed quartz, which contributes to intense dissolution of quartz in a glass phase. Figure 3 shows that curves 1, 2 and 5 for the cakes of the rocks studied exhibit a non-uniform course of thermal expansion. A rapid increase in TCLR in the range 600 – 700° C is typical of compositions with a minimum quartz concentration and high microcline and plagioclase concentrations which form a less viscous liquid phase. Large quantities of quartz in rocks are responsible for the rectilinear course of the TCLR curves (3, 4, 6 and 7). A minimum TCLE value (7.10⊙10⁻⁶ 1/deg) in the range 600-700° C is observed in cake 7 with a minimum quartz concentration (44.0 mass.%), and a maximum TCLR value (8.9⊙10⁻⁶ 1/deg) in cake 1 with a quartz concentration of 2.5 mass. %.

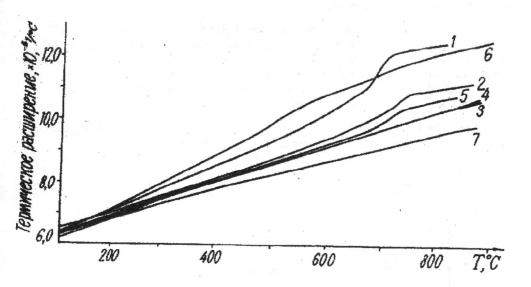

Fig. 3. Temperature dependence of the thermal coefficient of linear expansion: (numbers as in Table 1).

According to the literature [5] the dependence of TCLR on quartz concentration is due to a difference in the density of minerals. It is known that as mineral density increases, TCLR decreases. As the density of quartz (2.65 g/cm²) is higher than that of albite (2.61 g/cm²) and that of microcline (2.55 g/cm²), low TCLR values in cakes with high quartz concentrations can well be expected. The dependence of TCLR on microcline-plagioclase ratio shows a more complex pattern. As quartz concentration rises with a decline in feldspar concentration [1], also observed in Roza-Lambi quartz porphyry, the porcelain burning expansion interval will expand to 100 ° C without deformation of products upon burning.

Quartz has thus been shown to decrease the TCLR of feldspathic rocks, as indicated by the absence of rapid changes in the curves (Fig.3) in the range 600 – 700 ° C. High concentrations of feldspathic minerals (microcline, plagioclase - 95 – 97.5 mass. %), which form a less viscous glass phase, increase the TCLR of feldspathic rocks.

4. pH of feldspar suspension

pH is a parameter showing the acidity of feldspathic rocks and the engineering properties of clay slurry (viscosity, the density of clay slurry, tixotropy for molding ceramic products. As pH rises, the tixotropy of the clay slurry increases. Interacting with water, feldspathic minerals change the acidity of ceramic shlicker, depending on their constituent cations and anions. The optimum viscosity of clay slurry for refractory products is achieved at pH 5.5 and that for porcelain at pH 7.5–9. According to GOST 21119.3-91 for feldspathic fillers in varnish and paint production, pH is 6-9.

The pH of suspensions was measured using the potentiometer method on an I-120.1-type ionometer. Suspensions were prepared from из finely dispersed samples (grain size 0.063 mm) at solid:fluid ratio of 1:2.5 in accordance with the procedure described in [6,7]. The results of measurements of the pH of feldspar rock suspensions are presented in Table 1, and Figure 4 shows the dependence of pH on total microcline and plagioclase concentration in the rock.

Our results have shown that a common pattern of variations in the pH of feldspathic rock suspensions is a rise in pH with increasing alkaline oxide concentration.

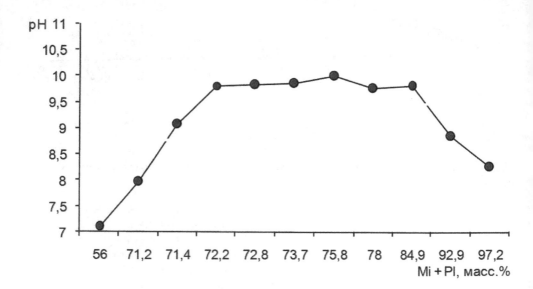

Fig. 4. Dependence of variations in pH on microcline and plagioclase.

Table 1 shows that suspensions of Khetolambino and Lupikko microcline pegmatites have the highest pH values: 9.67 and 9.82, respectively. Potassic halleflinta, a nonconventional type of feldspathic raw material, has a high pH (10.0). Comparison of the pH of quartz-feldspathic rock suspensions has shown that large quantities of quartz (pH of quartz is 6.4-6.9) decrease their pH considerably, as in Roza-Lambi volcanics (pH 7.1).

Figure 4 shows that rocks that contain a total microcline and plagioclase concentration of 75.8–78.0 mass. % and quartz concentration of 22.0–24.2 mass. % have the highest pH.

Thus, the pH of feldspar suspensions depends on total microcline and plagioclase concentrations and quartz concentration in the rock. The pH of Elisenvaara syenite is 8.25 is consistent with feldspathic raw material according to GOST 21119.3-91 (pH 6-9) and Kemio feldspar from Finland (pH 8.3).

5. Conclusions

The electrical properties ε , lg ρ, tg δ of the cakes of nonconventional types of feldspathic rocks and conventional types, such as pegmatite, are basically affected by alkaline oxides and quartz. Dielectrical permeability depends more on microcline and quartz concentrations in rocks. The angle tangent of dielectrical loss and electrical resistance depend on total microcline and plagioclase concentration. Elisenvaara syenites, which contain high microcline and plagioclase concentrations (97 mass.%), typically show lower tg δ values (0.025) and higher electrical resistance (2.0$\odot$10^{-12} Ohm$\odot$cm) than other rocks. Their TCLR is consistent with that of microcline pegmatite. The pH of the suspensions of nonconventional feldspathic raw material depends on total microcline and plagioclase concentration and quartz concentration. Elisenvaara alkaline syenites are a promising feldspathic raw material for electrical porcelain production.

6. References

[1] Avgustinnik, A.I. Ceramics. L.: Stroyizdat, 1975. 590 p.
[2] Avgustinnik, A.I., Kholodok, N.I, Golod, M.I. et al. On the thermal-physical properties of feldspathic minerals from pegmatite veins. // Geophysical study of the eastern Baltic Shield. Petrozavodsk, 1980. P.108-117.
[3] Ilyina, V.P. Technological evaluation of feldspathic rocks from the Louhi district // Geological and technological evaluation of minerals and rocks from the Republic of Karelia and some regions of the European continent. Petrozavodsk, 1997. P.57-59.
[4] Golod, M.I., Grodnitsky, L.L., Klabukov, B.N. On the dielectrical permeability of plagioclase from pegmatite veins of the Kola Peninsula// Indicator minerals of the characteristics of their host environment. L., "Наука",1975. P. 47-49.
[5] Kholodok, N.I., Golod, M.I., Popova, I.A., Klabukov, B.N. On the dielectrical permeability of potassic feldspars from pegmatite veins // Pegmatites from Karelia and the Kola Peninsula. Petrozavodsk, 1977. P.160-164.
[6] Karyukina, V.N. Determination of minerals from the pH of their suspension // Modern methods for mineralogical study of rocks, ores and minerals. M.: Gosgeotekhizdat, 1957. P.208-230.

[7] Kopchenova, E.V., Karyukina, V.N. Estimation of the pH of mineral suspension and the implication of this index in mineralogical studies // Modern methods for mineralogical study. M.: Nedra, 1969. P.148-155.

9

Sintering of Transparent Conductive Oxides: From Oxide Ceramic Powders to Advanced Optoelectronic Materials

Guido Falk

Saarland University, Chair Powder Technology of Glass
and Ceramics, Saarbruecken,
Germany

1. Introduction

Since Cadmium thin films have been sputtered and thermally oxidized for the first time in 1907 (Chopra, et al., 1983) the technological innovation in transparent conductive oxides (TCO) has developed rapidly and substantially during the last years. Indium-tin-oxide (In:90:Sn:10, ITO) is today the most important TCO material due to the potential combinations of high electrical conductivity in the range of 10^4 S/cm and high transparency in the visible range of 90 % with a layer thickness of 100 nm.

Today TCO thin films are processed by chemical or physical vapour deposition, especially vacuum based sputtering and evaporation processes. Due to these elaborate and complex manufacturing, especially for large-size TCO applications direct structuring of TCO layers by printing, sol-gel coating and other powder- and paste-based manufacturing processes have been the subject of many investigations (Hyatt, 1989, Straue, et al., 2009).

These as-processed TCO materials serve as transparent electrodes in liquid crystal displays, thin-film electroluminescence displays, electrochromic displays, transparent conductive coatings of highly sensitive radiation detectors, ferroelectric photoconductors and memory devices, transparent conductive oxidic films as gate-electrodes for injection and charge coupled devices and are used in products as flat panel displays, touchscreens, organic light emitting diodes, electroluminescence lamps as well as numerous components of solar technology.

Resulting from the increased indium consumption and the expected supply gap of indium raw materials within the next decades (Carlin, 2007) an increasing priority is attributed to the development of suitable substitute materials, whereas currently aluminium doped zinc oxide (ZnO:Al, AZO) have consistently attached the highest importance (Ellmer, et al., 2008). Beside the TCO applications mentioned above AZO is used in highly promising developments such as blue and ultraviolet lasers, components of improved light amplification of GaN-based LEDs, transparent thin film transistors, photo detectors, varistors, catalysts and optical current transformers.

Sputtered Al:ZnO thin films have already been used in commercially available flat panel displays und thin layer solar cells with an electrical resistance of 1-3 ·10⁻⁴ Ω cm at aluminium doping concentrations between 1.6 to 3.2 at.-% (Anders, et al., 2010).

Among others current research is focussing on sputtering process stability based on ITO target materials, on AZO target materials with improved electrical conductivity, thermal and mechanical stability as well as highest transparency of TCO thin layers by enhanced optimization of chemical composition, microstructure tailoring and advanced sintering and densification methodologies of TCO target materials.

2. Microstructure-property relationship of TCO materials

2.1 Nodule formation

Nodules, also called „black growths" or „black crystals", are conical defects formed during the sputtering process on the surface of the target material. The influence of nucleation on nucleus growth of nodules was investigated by camera monitoring (Schlott, et al., 1996). It was found, that especially impurities originating from the sputtering process, i.e. SiO_2 or Al_2O_3 particles , were collected on the target material and act as nucleus in the nodule formation process.

There are also other important impurities in the form of inclusions that originates from the target manufacturing process. Target materials that happen to feature the increased densification, are more suitable for sputtering processes since they exhibit fewer nodule failures (Schlott, et al., 1995, Gehman, et al., 1992). The shape of nodules has been investigated by SEM analysis and partly a distinct peak is observed, partly a flattened conical shape is formed (Schlott, et al., 1996). From time to time spherical globule like particles are observed at the nodules peak being identified as SiO_2 and Al_2O_3. In this case the particles could be assigned to impurities of the sputtering process being directly related to the nodule formation process. It was found by further detailed investigations that the external layer of the nodules contains pure ITO phase, the thickness of the layers being up to several tens of microns. Additionally these layers containing fewer oxygen concentrations compared to the bulk target material are supposed to be formed by self-sputtering. Self-sputtering means that the target itself was coated by the sputtering process as it is shown schematically in Figure 1. The oxygen deficiency can be explained by the fact that the atmosphere nearby the target surface contains less oxygen.

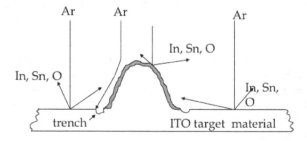

Fig. 1. Cross section scheme of a nodule formed on the surface of an ITO sputtering target according to (Schlott, et al., 1996).

Nodules have to be avoided, since they modify the sputtering process and it is therefore necessary to interrupt the process and to clean or even to exchange the target material. Nodule formation is a severe problem since nodules show a reduced sputtering voltage compared to the surrounding material. Thereby the nodules prevent the sputtering of the material being covered by the nodule layers. Furthermore when the nodule formation is avoided, arc discharge is not needed and the sputtering process is excecuted at increased sputtering voltages enabling to operate the facility at higher efficiencies (Nadaud, et al., 1995). The nodule formation was observed for metallic indium-tin targets during the reactive sputtering process (Schlott, et al., 1996) and also for sputtering of oxide ceramic ITO targets (Schlott, et al., 1995). The elimination of inclusions and metallic phases is predominant for the effective avaoidance of nodule formation (see Figure 2).

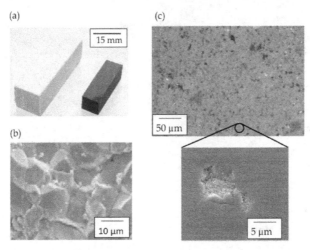

Fig. 2. Cold isostatically pressed ITO body (a, left) and sintered ITO body with a shrinkage of 15 % (a, right), SEM topography of fractured ITO surface after sintering with a mean grain size of 25 μm (b) and outbreak (dark) and segregations (bright) analysed by SEM at polished ITO surfaces (c).

A trouble-free microstructure is important to prevent the chipping (Schlott, et al., 1996). Impurties and metallic InSn eutectica have to be avoided in order to guarantee sufficient thin film qualities (Schlott, et al., 1996).

Once being formed the nodules grow contineously during the sputtering process. They do not dissolve unless they explode due to thermal stresses or due to the power of micro arc discharge effects (Schlott, et al., 1996). If that is done the desintegrated nodule particles are scattered and form nucleus of new nodules or they form small holes, so called pinholes, in the growing layer contributing to a significant quality degradation of the sputtered thin film (Kukla, et al., 1998).

Nodule formation is observed particularly frequently when the target material consists of several tiles being composed to enlarge the target surface. The split between the tiles are considered as collecting sites for impurites and dust. Furthermore the split could be coated by self-sputtering. If an Ar atom is smashed onto these coated splits, the scattered particles composed of In, Sn and O are able to be deposited onto the sputtered thin films (Schlott, et al., 1996). A key component for the suppression of undesired nodule formation is the basic understanding of most important ITO characteristics as well as specific sintering techniques and its effectiveness to tailor distinct microstructural properties of TCO target materials. In the following sections an overview of the state of the art and future trends of the related topics are given.

2.2 The system ITO and related defect structures

Enoki (Enoki, et al., 1991) proposed the following In_2O_3-SnO_2 phase diagram shown in Figure 3.

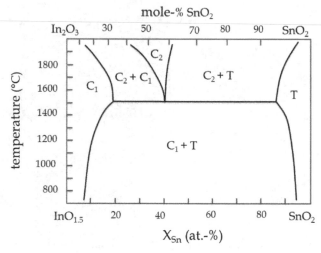

Fig. 3. Phase diagram of the pseudo-binary system In_2O_3-SnO_2 (χ_{Sn} = atom concen-tration of tin (%)).

Hereby the abbriviations C_1 and C_2 represent the cubic ITO phase and the orthorhombic intermediate respectively, and T is the rutile type SnO_2. It was found that the C_2 phase is formed in the concentration range between 47.9 and 59.3 mole-% Sn at temperatures exceeding 1573 K.

Morover it was oberserved that the intermediate (C_2) falls apart into In_2O_3 (C_1) and the T-phase according to the following eutectic reaction:

$$C_2 \leftrightarrow C_1 + T \tag{1}$$

Finally it was observed that the solubility limit of SnO_2 in In_2O_3 phase is between about 12.4 to 15.0 mole-% which is independent of the temperature (Enoki, et al., 1991). For some basic understanding it is helpful to consider the crystal structure of solid In_2O_3. The C type rare earth sesquioxide-or bixbyite crystal structure of Indium oxide is a variation of the cubically body-centred crystal structure (Marezio, 1966).

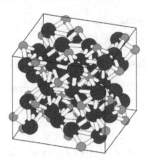

Fig. 4. Cubic In_2O_3 unit cell.

The unit cell has 80 atoms or 16 molecular weights (Warschkow, et al., 2003) and has two unequal cation positions (Granqvist & Hultaker, 2002). The lattice constant is 10.117 ± 0.001 Å (Marezio, 1966). The space group is referred to as Ia^3 or Th^7. Pure In_2O_3 has a density of 7.17 g/cm^3 (Bates, et al., 1986) and the theoretical density of the cubic structure is 7.12 g/cm^3 (Marezio, 1966). The electron subshell configuration of both atoms indium and oxygen is as follows according to (Chopra, et al., 1983):

$$O: \quad 1s^2 \mid 2s^2\, 2p^4 \text{- -} \mid \tag{2}$$

$$In: \quad 1s^2 \mid 2s^2\, 2p^6 \mid 3s^2\, 3p^6\, 3d^{10} \mid 4s^2\, 4p^6\, 4d^{10} \mid 5s^2\, 5p \tag{3}$$

The oxygen atoms need two more p-electrons to reach a stable 8-electron configuration. The indium atoms have three electrons in addition to a stable electron configuration. As a result the stoichiometry of the oxide is In_2O_3 resulting in a transition of electrons from In to O and a crystal structure with In^{3+} and O^{2-} ions in the lattice (Mayr, 1998). The unit cell has two unequal In-positions, the first space-diagonally (d position) and the other surface-diagonally (b position). With the d positions the oxygen atoms are located in the corners of an slightly distorted cube with two space-diagonal empty sites. In the second cube the oxygen atoms are located in a slightly differently distorted cube with two surface-diagonal empty sites. The characteristics of the defect structure are determined by this special arrangement in atoms (Hwang, et al., 2000). In_2O_3 exhibit a cubic bixbyite structure (Marezio, 1966), SnO_2 in contrast has a tetragonal structure (see Fig. 4) similarly to the rutile structure (Enoki, et al., 1991). The density of SnO_2 is 6.95 g/cm^3 (Bates, et al., 1986).

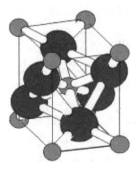

Fig. 5. Tetragonal SnO_2 unit cell.

Indium atoms are incorporated in the lattice as In^{3+} and the tin atoms as Sn^{2+}. The tin atoms have the following electron configuration (Mayr, 1998):

$$Sn : 1s^2 \mid 2s^2\,2p^6 \mid 3s^2\,3p^6\,3d^{10} \mid 4s^2\,4p^6\,4d^{10} \mid 5s^2\,5p^2 \qquad (4)$$

It was shown by means of X-ray diffraction analysis (Frank & Köstlin, 1982) that the cubic In_2O_3 structure is preserved by doping with SnO_2 up to the solubility limit of Sn in In_2O_3. The exact solubility limit of Sn in In_2O_3 is not exactly known and varies between 6 ± 2 at.-% of Sn. Up to this concentration every tin atom is substituted by an indium atom. The solubility in thin layers is even higher (Hwang, et al., 2000). The maximum solubility of Indium in the SnO_2 lattice is as low as 1 at.-%. Thereby Sn^{4+} ions are substituted by In^{3+} ions significantly decreasing the electrical conductivity. The ion radius of the Sn^{4+} is 0.71 Å and should lead to a linear reduction of the latticed constant with increasing doping of Sn^{4+} according to the Vegard law (Nadaud, et al., 1998) since the ion radius of In^{3+} is 0.81 Å. However, this is not observed. Udawatte (Udawatte, et al., 2000) reports on a maximum lattice constant reached at 5 mole-% of Sn content dropping below the maximum solubility of 6 mole-% of Sn in the In_2O_3 lattice reported by Nadaud (Nadaud, et al., 1998). These authors have calculated a lattice constant of 10.1247 Å at the maxium solubility limit of Sn compared to a lattice constant of 10.1195 Å that is observed for pure In_2O_3. The presence of Sn strongly changes the behaviour of the oxygen ions. Due to Sn doping the general distance between oxygen and cation increases, but the distance between oxygen and Sn decreases (Nadaud, et al., 1998). Typical phase modifications of indium-tin oxide are listed in the following table.

phase	unit cell	Structure	lattice constant (Å)	density (g/cm³)
In_2O_3	cubic	Bixbyite	10.117	7.17
SnO_2	tetragonal	Rutile	-	6.95
$In_4Sn_3O_{12}$	rhomboedric	Fluorite	-	7.30

Table 1. ITO phases, unit cells, structures, lattice constants and theoretical desities.

The formation of Indium oxide and the subsequent reaction contributing, inter alia , to the defect structure can be described by Kröger-Vink notation (Rahaman, 1995) according to the following structure elements of the chemical reactions (see Table 2).

abbreviaton		subscript term		superscript term	
V	vacancy	In	In lattic site	x	neutral
In	In atom	O	O lattice site	•	positively charged
O	O atom	i	Interstitial	′	negatively charged
e	electron	g	gas phase		
h	hole				

Table 2. Structure elements of chemical reactions according to Kröger-Vink notation.

The formation of Indium oxide is described according to the following reaction equation:

$$In_2O_3 = 2In^x{}_{In} + 3O^x{}_O \leftrightarrow 2In_g + 3/2O_g \qquad (5)$$

In parallel to this reaction the formation of gaseous In_2O can occur:

$$2In^g + (1/2O_2)_g \leftrightarrow (In_2O)_g \qquad (6)$$

Furthermore In_2O_3 is produced from gaseous In_2O^g in oxygen atmospheres according to the following reaction:.

$$In_2O_3 \leftrightarrow 2In^x{}_{In} + 3O^x{}_O \leftrightarrow (In_2O)_g + (O_2)_g \qquad (7)$$

Taking account of mass action law it is obvious that the oxygen partial pressure controles the number of oxygen vacancies $V^x{}_O$. The number of charge carriers in In_2O_3 is thus depending very much on oxygen partial pressure. Theoretically the number of charge carriers would increase at decreasing oxygen partial pressures associated with an increase of electrical conductivity (Mayr, 1998). However, the mechanismen is more complicated in reality since scattering mechanisms can occur consequently decreasing the charge carrier mobility and ion conductivity. More detailed information about scattering mechanisms are given in (Chopra, et al., 1983).

Oxygen vacancy concentration is a function of the defect structure of indium tin oxide (Freeman, et al., 2000). At the beginning first oxygen vacancies start to form as soon as the oxygen atoms leave their interstitial lattice sites $((2Sn^{\bullet}{}_{In}O_i'')^x)$ and are transformed to gaseous oxygen phase $(1/2O_{2g})$ (Freeman, et al., 2000). The following equation describes the formation of oxygen vacancies and the release of two electrons:

$$(2Sn^{\bullet}{}_{In}O_i'')^x \leftrightarrow 2Sn^{\bullet}{}_{In} + (1/2O_2)_g + 2e' \qquad (8)$$

Only under reducing conditions $(\sim pO_2{}^{-1/8})$ oxygen vacancies are formed by diffusing into the bulk from their former lattice sites $(O^x{}_o)$ (Hwang, et al., 2000). Under extreme reducing conditions $(pO_2 \sim 10^{-14} atm)$ non-reducing defects are formed such as $(2Sn^{\bullet}{}_{In}3O_OO_i'')^x$ (González, et al., 2001). The equation (9) decribes the transformation of gaseous species whereas two electrons for any oxygen vacancy. Equation (10) shows that the number of oxygen vacancies $(V^{\bullet\bullet}{}_o)$ formed and the number of doped lattice sites $(n = [D^{\bullet}{}_{In}])$ is a function of oxygen partial pressure $(pO_2{}^{1/2})$. This relation is described by the equilibrium constant K:

$$O^x{}_o \leftrightarrow (1/2O_2)_g + V^{\bullet\bullet}{}_o + 2e' \qquad (9)$$

$$K = pO_2{}^{1/2}[V^{\bullet\bullet}{}_o]n^2 \qquad (10)$$

If one oxygen atom is missing in the unit cell the valance electrons of the surroun-ding atoms have a reduced ionisation energy, which then is provided by thermal lattice vibrations. These electrons are in a quasi-free state and act as conduction electrons. That means that oxygen vacancies provide electrons for the conduction bands (Mayr, 1998). Even in the undoped state small oxygen deficencies can be detected. In this case oxygen vacancies appear in reduced concentrations compared to other defect structures (Hwang, et al., 2000).

system In_2O_3:Sn	
vacancy formation & self-compensation	$2In_{In}{}^x + 2SnO_2 \leftrightarrow (2Sn^{\bullet}{}_{In}O_i{}'')^x + In_2O_3$ (11)
donator formation	$(2Sn^{\bullet}{}_{In}O_i{}'')^x \leftrightarrow 2Sn^{\bullet}{}_{In} + (1/2O_2)_g + 2e'$ (12)
	$O^x{}_o \leftrightarrow (1/2O_2)_g + V^{\bullet\bullet}{}_o + 2e'$ (13)
	$2In_{In}{}^x + 2SnO_2 \leftrightarrow 2Sn^{\bullet}{}_{In} + 2e' + In_2O_3 + (1/2O_2)_g$ (14)

Table 3. Overview of reactions on formation and self-compensation of vacancies as well as formation of donators of the system In_2O_3:Sn according to Kröger-Vink notation.

Nadaud and co-workers investiated oxygen concentration of bulk-ITO by neutron diffraction and Rietveld analysis (Nadaud, et al., 1998). After sintering at 1400 °C in reducing nitrogen atmospheres stoiciometric oxygen concentrations were detected for both undoped and 6 at.-% Sn doped In_2O_3. On the other hand sintering in oxygen atmospheres resulted in bulk oxygen excess of about 3 %. This could be explained by the stoiciometry of the neutral $(2Sn^{\bullet}{}_{In}O'{}_i)^x$ complexes. Another explanation is the formation of large Sn-based oxygen complexes being difficult to get reduced in the intermediate temperature regime. The same authors investigated these large complexes by Mössbauer, EXAFS and neutron diffraction.

ITO Mössbauer- and EXAFS data von ITO reveal a relaxation of the Sn-O shell similar to the observed relaxation in Sn rhich $In_4Sn_3O_{12}$ (Nadaud, et al., 1998). The analytical data allow to extend the explanation of inefficacy doping above 6 at.-% Sn to the effect that Sn atoms are incorporated at cation sites where they are inactive.

The scientific community is controversial discussing the precipitation of the rhomboedric phase $In_4Sn_3O_{12}$ at Sn concentrations exceeding 6 at.-% Sn. The $In_4Sn_3O_{12}$ structure has been analysed by neutron diffraction experiments and it was ascertaiend that the structure is similar to the Fluorite structure and that a sixfold occupation of M1-sites by Sn cations and sevenfold occupation of M2-sites by In- and Sn cations occurs (Nadaud, et al., 1998). The following figure shows the unit cell of the rhomboedric $In_4Sn_3O_{12}$ phase. The lattice constant is 6.2071 Å and the mismatch angle ø is 99.29°.

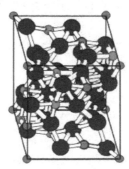

Fig. 6. Unit cell of the $In_4Sn_3O_{12}$ phase.

The atoms of the rhomboedric cells are more densely packed compared to the cubic structure. The rhomoedric phase was fist discovered by Bates et. al. and the density was calculated to 7.303 g/cm^3 (Bates, et al., 1986). It was found by X-ray diffraction experiments that the conformation is a densely packed $M'_mM''_nO_{3m}$ defect structure typically observed for the compositions Yb_7O_2 and Pr_7O_{12}. The In_2O_3 as well as SnO_2 solubility in the $In_4Sn_3O_{12}$-phase is limited (Bates, et al., 1986).

3. Synthesis of nano- and microcrystalline ITO ($Sn:In_2O_3$) and AZO ($Al:ZnO_2$) powders

3.1 ITO powder synthesis

In most cases indium-tin-oxide powders are synthesized by hydrothermal processes. Gel formation is based on the co-precipitation of $InCl_3 \cdot H_2O$ and $SnCl_2$ educts (Udawatte & Yanagisawa, 2000). The microcrystalline powder were homogeneous and reveal the composition of tin doped indium oxide [$In(OH)_3$:Sn] and tin doped indium hydroxide [InOOH:Sn]. Calcination of $In(OH)_3$:Sn at 300 °C resulted in cubic tin doped indium oxide [In_2O_3:Sn]. At calcination temperatures above 500 °C the InOOH:Sn phase is transformed to a solid solution of the formula ($In_2Sn_{1-x}O_{5-y}$). Both powders have been calcined in air atmosphere. It was found by Mössbauer analysis that the Sn^{4+}-ion coordination number is 8 because each tin atom has 8 neighbouring oxygen atoms which bear opposite charge. A similar synthesis scheme is presented in (Yanagisawa, et al., 2000). In this case indium-tin-oxide has been synthesized from a In-Sn-hydrogel. The hydrothermal treatment of the gel at 300 °C resulted in the formation of InOOH:Sn with mean particle sizes in the range of 80 nm. The subsequent calcination of the product at different calcination temperatures led to different microstructural vacancy configurations of indium tin oxide solid solutions. Calcination at 700 °C lead to a powder with primary particles sizes in the range of about 160 nm. Another approach is the processing of aqueous $In(NO_3)_3 \cdot H_2O$ solutions, subsequent heating and calcination at 500 °C in order to achieve nanosized ITO powders (Sorescu, et al., 2004). The solution of metallic tin and indium in HCl is precipitated by [NH_4OH] (Nam, et al., 2001) and the resulting indium-tin-hydroxide gel is dried, grinded and calcined at 600 °C.

The processing of tin doped indium oxide crystalites from a direct indium-tin smelting enriched with oxygen is described in (Frank, et al., 1976). The hetero-geneous nucleation process was initiated by small In_2O_3 crystalites. Different compositions and concentration ranges of the smelting process resulted in different tin doping concentrations. Indium tin oxide synthesis by a chemical transport process is reported in (Werner, et al., 1996). Starting from metallic indium and tin and dissolution of the educts in nitric acid the solution is dried and calcined at 900 °C to achieve In_2O_3 and SnO_2. The mixture of both oxides were doped with transport media iodine, or sulfur, or chlorine. Indium oxide crystals doped with 8.2 mole-% tin and tin oxide crystals doped with 2.4 mole-% indium were attained. These powders have been characterized and the synthesis reactions based on chlorine transport media have been thermodynamically modeled (Patzke, et al., 2000).

In case the seperate oxides are provided as educt materials for the synthesis reaction, indium doped tin oxide, as reported in (Nadaud, et al., 1994) could be processed. In this case In_2O_3 und SnO_2 powder (purity of 99,99 wt.-%) have been mixed in ethanol and calcined at

1380 °C for theree hours. Colouration from yellow to green and XRD peaks of indium-tin-oxide without exception proved the complete transforma-tion to indium tin oxide.

An alternative ITO synthesis process is given in (Stenger, et al., 1999). Here molten metallic indium tin alloy has been reacted with oxygen in a plasma arc furnace. The product is subsequently quenched by a gas stream at cooling rates between 10^5 K/s to 10^8 K/s to a final temperature between +50 °C and +400 °C. Thereby mixed indium tin oxide powders were produced with specific surface area of 3 m²/g at maximum. As the powder tends to formation of large agglomerates the size of different ramifications has been analysed by electron microscopy and documented as primary particle size being in the range of 0.03 μm up to 1.0 μm. Hot melt spraying of indium tin melts in oxydizing atmospheres and subsequent quenching is also suitable to process indium tin oxide powders. The agglomerated particle are porous and reveal particle sizes up to 170 μm (see Figure 4).

Compared to this the commercial powder ZY-ITO (Dow Chemical Co. Ltd.) is composed of a mixture of 90 wt.-% In_2O_3 and 10 wt.-% SnO_2, with a specific surface area of 32.0 m²/g and a mean primary particle size in the range of 3.4 μm.

3.2 AZO powder synthesis

Although various studies of synthesis, characterization and application of Al:ZnO thin layer systems are given in the literature (Stanciu, et al., 2001, Selmi, et al., 2008), there are only a few investigations referring to synthesis and processing of nanosized Al:ZnO particles. Flame spray pyrolysis of liquid Al- ad Zn precursors led to uncontrolled Al:ZnO stoichiometries and segregations of the spinel and wurzite phase combined with evaporation of the metallic Zn phase (Kim, 2008). Vapour condensation (Strachowski, et al., 2007) and spray pyrolysis (Hsiao, et al., 2007) as well as low-pressure spray pyrolysis (Hidayat, et al., 2008) resulted in highly agglomerated particles exhibiting poor redispersing characteristics.

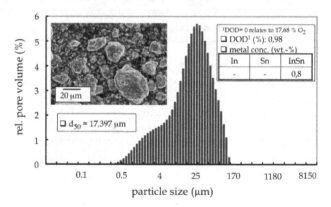

Fig. 7. Indium tin oxide powder produced by hot melt spraying and subsequent quenching: Pore volume of the powder as a function of particle size.

The processing of suitable AZO particles for optoelectronic applications is therefore still challenging. Several alternative processing routes have been elaborated with limited success. AZO particles synthesized by co-precipitation (Nayak, et al., 2007, Aimable, et al., 2010, Shui, et al., 2009) and solvothermal decomposition (Thu & Maenosono, 2010) are characterized by wide particle size distributions and heterogeneous Al doping concentrations of the microstructure. Hydrothermally (Piticescu, et al., 2006) and sol-gel based processing (Chen, et al., 2008) of AZO particles resulted in particles of high agglomeration degree and mean particles diameters considerably higher than 30 nm and undefined redispersing behaviour. The sintering and densification kinetics of Al:ZnO green bodies is slowed down with increasing Al doping concentrations and the grain growth effects are significantly reduced (Han, et al., 2001, Hynes, et al., 2002). Information about processability of synthesized AZO particles within technical and functional ceramic processing chains is still missing.

4. Sintering of TCO materials

4.1 Sintering and Degree of Oxygen Deficiency (DOD)

The degree of oxygen deficiency (DOD) is referred to oxygen deficient in relation to oxygen concentration of the fully oxidized powder (Schlott, et al., 1995). In (Weigert, et al., 1992) the degree of oxygen deficiency is defined as follows:

$$DOD = (a-b) / (a-c) \qquad (15)$$

Here a is the oxygen concentration of stoiciometric oxidic compound, b is the oxygen concentration of the partially reduced oxidic compound and c is the oxygen concentration of the completely reduced metallic compound. The degree of oxygen deficiency is closely related to the oxygen vacancies contributing in parts to the electrical conductivity of ITO. Oxygen release is caused by thermolysis of the ITO powder and oxygen uptake is observed during cooling. The arrangement of oxygen vacancies is likely to be closely connected to the powder synthesis.

Thermal treatment in air leads to light-green colouring of the powder and thermal treatment in reducing atmospheres results in black colouring of the powder. Sputtering is causing a yellow colour of the target (Otsuka-Matsua-Yao, et al., 1997). After sintering the powder can also change colour to dark green or black(Udawatte, et al., 2000). The thermal history of the powder is of significant importance since the powder characteristics are closely correlated with the target quality having a profound impact on the characteristics of the sputtered thin layers (Weigert, et al., 1992).

Evidence proves that oxygen release and uptake of ITO follows a chemical hysteresis (Otsuka-Matsua-Yao, et al., 1997). In repeated heating and cooling cycles different quantities of oxygen are incorporated or released. According to the findings of experimental sintering investigations (Otsuka-Matsua-Yao, et al., 1997) the quantities of released oxygen below 773 K was marginal. A preliminar thermolysis at 1073 K was needed to release carbon monoxide absorbed on the surface. The authors came to the result, that the thermodynamic oxygen characteristics of ITO samples were changed during heating and cooling cycles. Hereby the total oxygen concentration is reduced during each cycle and the kinetics of oxygen release is slowed down cycle by cycle. The initiation temperaure of oxygen release was observed to increase on every iteration of heating and cooling cycle. It was assumed by the authors that there are several phases and several phase transformations inbetween these cycles causing the discontinuous oxygen release. The related phases exhibit similar chemical composition and resemble the CaF_2 structure. The phases are devided in coherent domaines and the surface energy could have a major impact on the *Gibb's free energy*. This means that chemical hysteresis appears and the oxygen concentration can not be determined explicitly as a function of the temperature and the oxygen partial pressure (Otsuka-Matsua-Yao, et al., 1997).

But there is evidence by experimental investigations that several microphases could act as barriers thereby preventing the oxygen release. These microstructures could originate from extended defects, these are sheared structures or domains, which diverge from each other in therms of chemical compositions on a very small scale. However the exact analysis and qualitiative and quantitative identification of these structures by X-ray diffraction is not possible since they are charcterized by the structure of the rare earth elements, as for example In_4SnO_8, $In_4Sn_2O_{10}$, In_2O_3 etc. The oxygen release could therefore depend on the interfacial energy between these microphases. In sputtering processes partially reduced indium-tin-oxide target materials are preferred in order to achieve thin film characteristics of minimum specific electric resistance and maximum sputtering efficiencies. Partially reduced targets reveal improved electrical and thermal characteristics as well as optimized densification.

The reduction of the powder or sintered body can be achieved by several methods, i.e. sintering in vacuum or reducing atmospheres (H_2, CO, H_2-Ar oder H_2-N_2). Alternatively the specimen can be hot pressed in graphite moulds or carbon releasing materials or the carbon releasing materials can be doped with the specimen (Weigert, et al., 1992). However under these reducing conditions the caculated oxygenn stoichiometry can not be accurately controlled. Target materials with reproducible and uniform degree of oxygen deficiency are preferred, since they guarantee superior sputtering efficiencies and reduced operation and coating times (Weigert, et al., 1992).

Furthermore the adding of reducing agents and ingredients causes normally the formation of metal particles of different sizes considerably larger than 50 µm (Schlott, et al., 1995). The microstructure appears spotty and targets processed by the powder qualities reveal lower fracture toughness characteristics. From this reason the target specimen shall be reduced below 1000 °C and preferably even below 800 °C. The subsequent densification of the powder and/or the sintered body is realized usually by hot isostatic pressing at temperatures above the melting point of the metal indium-tin-phases (Schlott, et al., 1995).

The hydrophilic characteristics of In_2O_3 should be noted. The water uptake from the surrounding atmosphere leads to the transformation of In_2O_3 to $In(OH)_3$. The water absorption capacity is increasing with increased specific surface area (Lippens, 1996).

Ceramic ITO target materials show low thermal conductivities compared to the metallic target materials and are therefore very susceptible to thermal stresses resulting form non-uniform heating during the sputtering process. It is therefore preferred to increase the thermal conductivity within certain limits in order to guarantee improved thermal shock resistance characteristics (Schlott, et al., 2001).

Based on those optimized target materials the sputtering process is able to be performed at higher specific powder densities without resulting in target desintegration due to formation of thermal gradients.

Improved thermal shock resistance characteristics also allow increased sputtering rates and shortened processing and operation times and consequently reduced production cost. At comparable sputtering rate increased thermal conductivity causes decreased target surface temperature and the decreased temperature gradient ensures a reduction of thermal stresses as well as increased reliability and target life time. This is especially true for target materials working as cathodes under pulsed sputtering operation conditions (Schlott, et al., 2001). In the state-of-the-art the thermal conductivity of ITO target materials is in the range of 7 – 10 W/mK.

In principle a high densification of the target materials is desirable, that means the density should exceed 95 % of the theoretical density in order to guarantee exceptable thermal conductivity characteristics. Usually the thermal conductivity could also be increased by partially reduction (Schlott, et al., 2001).

The theoretical melting point of indium-tin-oxide is 2223 °C. However above the temperature of 1600 °C ITO is evaporated due to a critical vapour pressure regime of the indium and tin metal phases (Vojnovich & Bratton, 1975). At a temperature of 1000 °C the vapour pressure of indium is $6.6 \cdot 10^{-5}$ bar and that of tin is $6.6 \cdot 10^{-7}$ bar (Nichols, 1982).

4.2 Specific TCO sintering techniques

4.2.1 Thermal vacuum degassing

Water absorbed in the green body has to be thermally desorbed in order to avoid crack formation and specimen desintegration after sintering and subsequent cooling. This is especially true for capsuled and hot isostatically pressed samples. The gas formation during sintering has to be prevented by preheating of green bodies within controlled degassing atmospheres as for example N_2 or Ar (Lippens, 1996). The thermal degassing process purifies the grain boundaries and is resulting in a reduction of specific surface area. Hereby the sintering activity is reduced and the densification has to be achieved by pressure assisted sintering techniques (Schlott, et al., 2001).

As referred to the previous section, thermal vacuum degassing exerts a strong influ-ence on the dgree of oxygen deficiency. Other studies (Falk, 2008) came to the result that vauum degassing and combined hot isostatic pressure sintering of capsuled ITO resulted in optimized sintering densities if the vacuum degassing time and temperature is correlated with a defect free micro structure, adopted concentration of free metal species as well as suitable degrees of oxygen deficiency (see Figure 8).

4.2.2 Pre-sintering

Based on the findings of Son (Son & Kim, 1998) it was shown that pressure assisted pre-sintering of In_2O_3 at 5 MPa could increase the densification rate significantly. Dilatometer experiments have proven that maximum densification rates were achieved at 1130 °C and the pressure was increased for another 5 MPa at this temperature. Sintering in air (1 atm) at 1500 °C, however, only leads to a desification of 76 % of the theoretical density.

In (Son & Kim, 1998) it was found, that the agglomerated green structure was transformed in a homogeneous polycrystalline microstructure at temperatures above 1070 °C and the agglomerates showes increased densification rates compared to the surrounding matrix leading to pore formation, so called interagglomerate pores, inbetween the matrix and the densified microstructure regimes. These pore formations could be effectively prevented by pressure assisted sintering and rearrangement of densified areas and by moving the pores to the surface area.

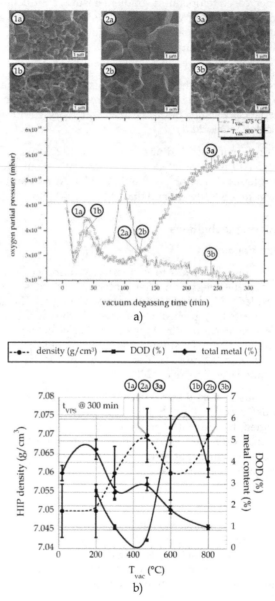

Fig. 8. Oxygen release and oxygen capturing and attained microstructures during vacuum degassing at 475 °C and 800 °C as a function of vacuum degassing time (a), HIP densities, free total metal content and degree of oxygen deficiency (DOD) as a function of vaccum degassing temperature of ITO sintered bodies (b) according to (Andersson, et al. 2005).

4.2.3 Sintering without additives

In (Stenger, et al., 1999) the avoidance of binders and/or dry pressing agents is proposd to prevent any contamination of the received ITO powder. Futhermore it is suggested to avoid the evaporation of gaseous species during pyrolysis of additives since these proceses are likely to reduce efficient pressure build up during hot isostatic pressing.

In (Udawatte, et al., 2000) the group reported about additive-free sintering of ITO powder compacts in air atmospheres. Starting from hydrothermally prepared and at 500 °C calcinated precursors ITO powders of the composition $In_2Sn_{1-x}O_{5-y}$ were attained. The authors have shown previously that pre-sintering of ITO and significant densification is achieved when the $In_2Sn_{1-x}O_5$ phase is transformed to cubic $In_{2-y}Sn_yO_3$ in the temperature range between 1000 °C and 1200 °C. Sintering necks were observed in this temperature range and by excceding the sintering temperature above 1250 °C sigificant grain growth was initiated. At 1300 °C a uniform grain size of 2 µm up to 3 µm and a sintering density of 65 % related to the theoretical density was observed.

The maximum densification was achieved at 1450 °C correlated with a mean grain size of about 7 µm. In this case triple grain boundary pores arised more and more frequently compared to intragranular pores.

The conclusion of these experimental investigations were that the sintering is activated mainly in the temperature range between 1300 °C and 1400 °C. The maximum density of about 92 % of the theoretical density was achieved after sintering at 1450 °C for three hours. Furthermore it was concluded that tin doping results in higher densification rates. In (Udawatte, et al., 2000) it is mentioned that tetragonal SnO_2 phase formation counteracts ITO densification.

In (Udawatte & Yanagisawa, 2001) small dry pressed powder compacts (diameter 10 mm, thickness 1.5 mm) have been sintered at 1400 °C for three hours. Taking the theoretical density of 7.106 g/cm³ as a basis a maximim density of 99.3 % of the theoretical density was achieved. The powder used had a mean particle size of about 80 nm.

Compared to conventional sintering elevated densities have been achieved by „spark plasma sintering" (SPS) (Takeuchi, et al., 2002). At a reduced dwell time of 5 minutes and high heating rates up to 50 K/min the SPS experiments resulted in considerably low sintering densities, probably due to inhomogeneous temperature distribution.

The sintering of cubic and rhomboedric nanosized ITO powders with mean particle sizes in the range of 50 to 100 nm were sintered up to 900 °C (Kim, et al., 2002) where the cubic phase was transformed. This transformation should theoretically results in a volume expansion of 2.1 % which was not observed since grain growth and pore formation were initiated. It was very complicated to eliminate these pores by subsequent sintering at elevated temperatures. The phase transformation promoted the active diffusion of atoms resulting in inhomogeneous grain growth with intragranular pore formation. It is therefore recommended to prevent phase transformation during sintering in order to achieve higher densification rates and more homogeneous microstructures (Kim, et al., 2002).

4.2.4 Sintering with additives

Vojnovich and co-workers (Vojnovich & Bratton, 1975) investigated the influence of imputities in terms of densification rates of In_2O_3 powders. Combinations of the impurities Si, Ca, Mg, Pb and Fe as well as the doping with Kaolin, Al_2O_3 and SiO_2 have been experimentally studied. It was found that the impurities forming liquid phases due to eutectic reactions even contribute to increased densification rates as for example by doping with Ca, Si and Mg as well as Kaolin. Impurities leading to low melting eutectic phase formations favour liquid phase sintering conditions and resulting in higher density values (Vojnovich & Bratton, 1975). It was found that TiO_2 doping causes an increase of sintering density and limits grain growth. From a concentration of 0.5 wt.-% TiO_2 as sintering additive the phase In_2TiO_5 is precipitated at the grain boundary resulting in increased grain boundary diffusion at reduced diffusion activation energies (Nadaud, et al., 1997).

The influence of Si and Zr impurities on ITO sintering was ivestigated in (Nadaud, et al., 1994). These studies have been motivated by the successful ITO-doping with tetravalent Ti (Nadaud, et al., 1997) and the attempt to achieve even better results with alternative tetravalent doping additives. It came out that a sigificant influence of SiO_2 on ITO densification was not observed. Zirconia, however, has a negative influence on densification and increase specific electrical resistance of ITO ceramics at room temperature. The approach to incorporate sintering additives in order to increase densification is being questioned in (Schlott, et al., 1996), since it was found that imputities, i.e. TiO_2, presented as inclusions or segregations in the microsructure, could cause the formation of nodules.

4.3 Sintering behavior and TCO microstructure

Yanai and Nakamura (Yanai & Nakamura, 2003) found that SnO_2 aggregates in the powder could negatively influence the sintering of ITO and could consequently result in higher porosity of the sintered body. In this case the walls of the pores are covered by SnO_2 segregations. In case of finely distributed SnO_2 segregations decreasing porosity is achieved. And finally in (Hsueh & Jonghe, 1984) it was agreed upon that inhomogeneous sintering takes place in case of heterogeneous density and stress distribution. The experimental evidence of these facts was given by measurements of heterogeneous stress distribution along the grain boundaries causing gradients of the chemical potential and acting as a driving force of the material transport.

Investigations of the microstructure of target materials processed at different processing conditions have shown that finely distributed metallic phases led to improved densification and significant recrystallization of the matrix during hot isostatic pressing.

This means that the liquid metal phase could further enhance the densification although under normal operation condition the wetting of indium oxide and tin oxide by the metallic phases of indium and tin is not observed (Schlott, et al., 1995).

The recrystallization leads to higher densities and and an essantial increase of fracture toughness (Falk, 2007, Falk, 2009) . This is unexpected too, since for ceramic materials a reduction of grain size leads to improved mechanical properties. Parallel to these findings it was also oberserved that the morphology of partially reduced powders have a major impact on densification and mechanical stability (Schlott, et al., 1995).

Studies of the fracture surface have shown that the intergranular bonding was much higher in recrystallized microstructures. Hereby the metallic phases take over a crack arresting function.

Good quality targets are attained at DOD values in the area of 0.02 to 0.2. If the degree of oxygen deficiency is too low metallic phases are rare and the positive influence of these phases on densification and mechanical strength is neglectable.

If the degree of oxygen deficiency is too high large area metal segregations act as microstructural failures and cause decreasing mechanical strength and worse densification (Schlott, et al., 1995). These authors processed targets of two different powders incoporting metal segregations of mean diameter between 1 μm to 10 μm and < 200 μm. Grain growth effects as a function of the seggregation size was detected and it was conclued that grain growth was especially pronounced in the case of large seggregated microstructures and consequently the maximum sputtering efficiency could not be achieved due to temporarily devating arc discharge at the surface of the target material.

In case of slightly reduced target materials especially tin segregation have been observed and in the event of substantially reduced targets the metallic phases are indentified as InSn alloys. The In-Sn eutectic phase with 48.3 at.-% Sn is melting at 120 °C according to (Shunk, 1969).

4.4 Sintering behavior and electrical conductivity

The electrical conductivity of both pure indium oxide and pure tin oxide is a result of stoichiometry disturbance due to formation of oxygen vacancies. In case of In_2O_3 this structure can be described by the complex $In_2O_{3-x}(V_o)_xe'_{2x}$ (Mayr, 1998). By tin doping, having a higher valence number compared to indium, negative charge carriers are incorporated into the lattice contributing to an additional increase of electrical conductivity.

Freeman and co-workers (Freeman, et al., 2000) have calculated the theoretical distribution of energy bands in tin doped and undoped In_2O_3. In the case of tin doped indium oxide the s-band in the lower section of the conduction band is broadened. Consequently a high mobility of electrons is achieved explaining the high electrical conductivity. Further investigation of electronic ITO structure was elaborated in (Odaka, et al., 2001).

In general the electrical conductivity of Sn doped ITO is lower than theoretically predicted. It was observed that high Sn doping concentrations even reduce the electrical conductivity. The $In_4Sn_3O_{12}$ phase mentioned before is not or slightly electrically conductive (Nadaud, et al., 1998). This fact can be explained by the inactivity of Sn-cations located in M1-positions due to the ternary coordiation of the surrounding oxygen ions. This is similar to the structure of SnO_2 which is neutral. Enventhough In and Sn are located in close vicinity of the periodic system of elements the additional electron of the Sn atom causes a higher affinity to oxygen compared to indium.

The fact that Sn is deteriorating the electrical conductivity becomes more reasonable when high doping levels exceeding 9 at.-% Sn doping concentrations at increased oyxgene partial pressures (pO_2 = 1 atm) are applied as it was experimentally proven by Nadaud (Nadaud, et al., 1998).

With the exception of the $In_4Sn_3O_{12}$-phase formation there are mainly two reasons explaining the decrease of electrical conductivity of ITO as soon as doping concentrations exceeding limiting values of 6 at.-% of tin. First the electrical conductivity is reduced due to the formation of the neutral irreducable clusters (Frank & Köstlin, 1982) as for example $(2Sn^{\bullet}_{In}O_i)^x$ complex or the strongly attached and neutral $(Sn_2O_4)^x$ complex (Frank & Köstlin, 1982). Second the lattice is progressively distored as soon as the doping concentration are increased. The atoms are displaced from their original positions and In_2O_3 is similar to SnO_2 crystalline structure. Consequently Sn_2O_i''-cluster are formed acting as neutral lattice defects (Nadaud, et al., 1998). Furthermore these formed clusters are able to restrict doped charge carrier concentrations und decrease charge carrier mobility by causing „un-ionized impurity scattering" (Hwang, et al., 2000).

At increasing oxygen vacancy concentrations a compensation process is initiated resulting in a further decrease of the electrical conductivity according to the following equation (Mayr, 1998):

$$O_i'' + V_o \Leftrightarrow O^x_o \qquad (16)$$

The electrical characteristics of ITO –phases have been experimentally investigated by Bates and co-workers (Bates, et al., 1986). After sintering of ITO the cubic body centred In_2O_3 phase, being able to incorporate tin concentrations up to 20 mole-% by solid solution process, the rhomboedric $In_4Sn_3O_{12}$ as well as the tetragonal SnO_2 have been detected. It was found that the electrical conductivity is increasing with increased In_2O_3 phase content up to a phase concentration of about 30 mole-%, passing a constant conductivity level up to about 50 mole-% In_2O_3 and reaching a maximum conductivity level at In_2O_3 phase concentration of about 80 mole-%.

This maximum electrical conductivity is a factor of 20 to 25 times higher than the electrical conductivity of pure In_2O_3 (1.6 up to $2.7 \cdot 10^3$ /Ωcm) and a factor of 6 to 20 times higher than the electrical conductivity of $In_4Sn_3O_{12}$ (100 up to 300 /Ωcm). These values are clearly lower than those of thin layers which are in the range of up to 10^4 /Ωcm (Nadaud, et al., 1994).

Studies of the cyclical heating and subsequent cooling of 10 up to 70 mole-% In_2O_3 in air resulted in a reproducible hysteresis of the electrical conductivity and the thermoelectric power. These characteristics are connected to the formation of the high temperature $In_4Sn_3O_{12}$ phase according to the authors (Bates, et al., 1986).

It is also claimed that there are other phase transformations at elevated temperatures, contributing to additional explanations of the fluctuations in electrical conductivity. In continuous thermogravimetrical studies of ITO targets a hysteresis of oxygen uptake and release was observed from multiple cyclical heating and cooling in atmospheres with controlled oxygen partial pressure (Otsuka-Matsua-Yao, et al., 1997). The conclusion is that the microstructural transformation of ITO in the temperature range between 1273 K and

1773 K is correlated with the oxygen content. These results also indicate that the typical ITO characteristics not only depend on a specific oxygen concentration range since a hysteresis is present. Investigations of electrical In_2O_3 characteristics at elevated temperatures have been elaborated by De Wit (Wit, 1975). Hwang (Hwang, et al., 2000) proposed three different regimes of electrical conductivity as a function of oxygen partial pressure and tin concentration. These three regimes have been ascertained experimentally by electrical conductivity and thermoelectric power measurements. The thermoelectric power is a measurement for the thermal diffusion current which is achieved by a temperature gradient. The first regime is characterized by low oxygen partial pressure ($\sim pO_2^{-1/6}$) and low tin doping concentrations. The second regime is distinguished by mean oxygen partial pressure ($\sim pO_2^{0}$) and mean tin concentrations and the third regime by high oxygen partial pressure ($\sim pO_2^{-1/8}$) and high tin doping concentrations. The different doping concentrations result in different defect structures.

The electrical conductivity of bulk nano-ITO is significantly lower compared to the electrical conductivity of bulk μ-ITO. The explanation seems to be that the charge carrier density and the mobility of charge carriers is much lower in nano-ITO (Hwang, et al., 2000). Modeling of optical and electrical characteristics of ITO-thin films made of nano-ITO have been elaborated by Granqvist and co-workers (Ederth, et al., 2003).

5. Conclusions and future trends

Ceramic transparent conductive oxides are widely used for the processing of thin transparent conductive oxides films by vacuum sputtering techniques. These thin film layers are used in liquid crystal display technologies and various application fields such as energy conservation, information storage, electrophotography, electromagnetic radiation shielding and optoelectronic industry. In order to achieve maximum electrical and thermal conductivities and high sputtering efficiencies usually TCO target materials with distinct degrees of oxygen deficiencies are being used. The transparent semiconductor indium-tin-oxide with its high transmission for visible light, its high electrical conductivity and its strong plasma reflection in the near infrared is one of the most common transparent conductive materials. A simplified description of basic understanding of most important ITO characteristics are given and correlated to the desired microstructural properties of sintered TCO target materials.

Specific sintering techniques, i.e. hot isostatic pressing of vacuum pre-sintered, compacted and capsuled ITO bodies results in a distinct consolidation of microstructure and a homogenisation of ITO phase and thus to an increase in HIP sintered density close to the theoretical density. In order to achieve homogeneous microstructures with small mean grain sizes of the pure ITO phase elaborate demands on specific sintering methodologies and adopted sintering processing chains have to be applied. It is important to note that the characteristics of industrially available raw powders in terms of phase composition, particle size distribution, powder density, degree of oxygen deficiency and concentrations of free metal species such as In, Sn and InSn intermetallics could have considerable impact on the quality of the later ITO product. The assessment of superior powder quality in the framework of stream lined and rationalized powder synthesis and processing can therefore be seen as a way forward to realize further optimization of ITO target materials for sputtering applications. Facing the shortage of indium raw material resources it will be of increasing importance within the coming decades to develop suitable substitute TCO materials. At present aluminium doped zinc oxide (AZO) is one of the most promising candidates. Although the attempts of development of flexible transparent electrode coatings on the base of TCO nano powders for large scale applications have yet to yield any spectacular breakthroughs, new cost-effective niche applications, i.e. flexible displays, are likely to rise up and compete with the state-of-the-art sputtering thin film methodologies in the near future. It is therefore expected that new modern sintering techniques of nano particulate TCO materials, i.e. laser treatment, will be developed in order to significantly improve electrical conductivity of TCO particulate materials on polymer substrates.

6. References

Aimable, A., et al. (2010). Comparison of two innovative precipitation systems for ZnO and Al-doped ZnO nanoparticle synthesis. *Processing and Application of Ceramics*, Vol. 4, No. 3, pp. 107-114, ISSN 1820-6131

Anders, A., et al. (2010). High quality ZnO:Al transparent conducting oxide films synthesized by pulsed filtered cathodic arc deposition. *Thin Solid Films*, Vol. 518, No. pp. 3313-3319, ISSN 0040-6090

Andersson, L., et al. (2005). On a Correlation Between Chemical Hysteresis and Densification Behavior of SnO_2-In_2O_3 Powder Compacts *cfi/Ber. Dtsch. Keram. Ges.,* Vol. 13, No. pp. 208-211, ISSN 0173-9913

Bates, J. L., et al. (1986). Electrical conductivity, Seebeck coefficient, and structure of In_2O_3-SnO_2. *Am. Ceram. Soc. Bull.,* Vol. 65, No. 4, pp. 673-678, ISSN 0002-7812

Carlin, J. F. (2007). Indium, In: *Minerals Yearbook: Metals and Minerals,* J. H. DeYoung, J., pp. 78-79, U.S. Department of the Interiour, U.S. Geological Survey, ISBN 978-1-4113-3015-3, Washington

Chen, K. J., et al. (2008). The crystallization and physical properties of Al-doped ZnO nanoparticles. *Appl. Surf. Sci.,* Vol. 254, No. pp. 5791–5795, ISSN 0169-4332

Chopra, K. L., et al. (1983). Transparent conductors - a status review. *Thin solid films,* Vol. 102, No. pp. 1-46, ISSN 0040-6090

Ederth, J., et al. (2003). Indium tin oxide films made from nanoparticles; models for the optical and electrical properties. *Thin Solid Films,* Vol. 445, No. pp. 199-206, ISSN 0040-6090

Ellmer, K., et al. (2008). *Transparent Conductive Zinc Oxide,* Springer Verlag, ISBN 978-3-540-73611-9, Berlin, Heidelberg, New York

Enoki, H., et al. (1991). The intermediate compound in the In_2O_3-SnO_2 system. *J. Mater. Sci.,* Vol. 26, No. 15, pp. 4110-4115, ISSN 0022-2461

Falk, G. (2008). Densification of $(In_{0.9}Sn_{0.1})_2O_3$ by vacuum pre-sintering and hot isostatic pressing. *cfi/Ber. Dtsch. Keram. Ges.,* Vol. 85, No. 13, pp. 35-38, ISSN 0173-9913

Falk, G. (2009). Heißisostatisches Pressen nanokristalliner Oxidkeramiken, In: *Technische Keramische Werkstoffe,* Kriegesmann, J., pp. 1-37, HvB-Verlag, ISBN 978-3-938595-00-8, Ellerau

Falk, G. (2007). Optimierung von ITO-Targets durch Hippen. *Fortschrittsberichte der DKG: Verfahrenstechnik,* Vol. 21, No. 1, pp. 102-111, ISSN 0173-9913

Frank, G., et al. (1976). The solubilities of Sn in In$_2$O$_3$ and of In in SnO$_2$ crystals grown from Sn-In melts. *J. Cryst. Growth*, Vol. 36, No. 1, pp. 179-180, ISSN 0022-0248

Frank, G. & Köstlin, H. (1982). Electrical properties and defect model of tin-doped indium oxide layers. *Appl. Phys. A*, Vol. 27, No. pp. 197-206, ISSN 1432-0630

Freeman, A. J., et al. (2000). Chemical and thin-film strategies for new transparent conducting oxides. *MRS Bulletin*, Vol. No. August, pp. 45-51, ISSN 0883-7694

Gehman, B. L., et al. (1992). Influence of manufacturing process of indium tin oxide sputtering targets on sputtering behavior. *Thin Solid Films*, Vol. No. 220, pp. 333-336, ISSN 0040-6090

González, G. B., et al. (2001). Neutron diffraction study on the defect structure of indium-tin-oxide. *J. Appl. Phys.*, Vol. 89, No. 5, pp. 2550-2555, ISSN 0021-8979

Granqvist, C. G. & Hultaker, A. (2002). Transparent and conducting ITO films: New developments and applications. *Thin solid films*, Vol. 411, No. 2002, pp. 1-5, ISSN 0040-6090

Han, J., et al. (2001). Densification and grain growth of Al-doped ZnO. *J. Mater. Res.*, Vol. 16, No. 2, pp. 459-468, ISSN 2044-5326

Hidayat, D., et al. (2008). Single crystal ZnO:Al nanoparticles directly synthesized using low-pressure spray pyrolysis. *Mat. Sci. Eng. B*, Vol. 151, No. pp. 231–237, ISSN 0921-5107

Hsiao, K.-C., et al. (2007). Synthesis, characterization and photocatalytic property of nanostructured Al-doped ZnO powders prepared by spray pyrolysis. *Mat. Sci. Eng. A*, Vol. 447, No. pp. 71-76, ISSN 0921-5093

Hsueh, H. H. & Jonghe, L. C. D. (1984). Particle rotation in early sintering. *J. Am. Ceram. Soc.*, Vol. 67, No. 10, pp. C215-C217, ISSN 0002-7820

Hwang, J. H., et al. (2000). Point defects and electrical properties of Sn-doped In-based transparent conducting oxides. *Solid State Ionics*, Vol. 129, No. pp. 135-144, ISSN 0167-2738

Hyatt, E. P. (1989). Continuous Tape Casting for Small Volumes. *Ceramic Bulletin*, Vol. 68, No. 4, pp. 869-870, ISSN 0002-7812

Hynes, A. P., et al. (2002). Sintering and characterization of nanophase zinc oxide. *J. Am. Ceram. Soc.*, Vol. 85, No. 8, pp. 1979-1987, ISSN 0002-7820

Kim, B. C., et al. (2002). Effect of phase transformation on the densification of coprecipitated nanocrystalline indium tin oxide powders. *J. Am. Ceram. Soc.*, Vol. 85, No. 8, pp. 2083-2088, ISSN 0002-7820

Kim, M. (2008). Mixed-metal oxide nanopowders by liquid-feed flame spray pyrolysis (LF-FSP): Synthesis and processing of core-shell nanoparticles, Ph.D., The University of Michigan, Michigan

Kukla, R., et al. (1998). Sputtered ITO-layers: new approaches for high-quality, low-cost-production, *Proceedings of twelth international conference on vacuum web coating conference*, pp. 104-111, Reno, Nevada

Lippens, P. (1996). Integration of target manufacturing in the sputtering plant, *Proceedings of 39th Annual Technical Conference Proceedings*, pp. 424-430, Albuquerque, ISSN 0737-5921

Marezio, M. (1966). Refinement of the crystal structure of In_2O_3 at two wavelengths. *Acta Cryst.*, Vol. 20, No. pp. 723-728, ISSN 1600-5724

Mayr, M. (1998). High vacuum sputter roll coating: a new large-scale manufacturing technology for transparent conductive ITO layers, Leybold-Heraeus, Report

Nadaud, N., et al. (1995), Matériau pour cible de pulverisation cathodique, EU 0679731, Patent, Saint Gobain Vitrage International

Nadaud, N., et al. (1997). Titania as a sintering additive in indium oxide ceramics. *J. Am. Ceram. Soc.*, Vol. 80, No. 5, pp. 1208-1212, ISSN ISSN 0002-7820

Nadaud, N., et al. (1998). Structural studies of tin-doped indium oxide (ITO) and $In_4Sn_3O_{12}$. *J. Solid State Chem.*, Vol. 135, No. 1, pp. 140-148, ISSN 1095-726X

Nadaud, N., et al. (1994). Sintering and electrical properties of titania- and zirconia-containing In_2O_3-SnO_2 (ITO) ceramics. *J. Am. Ceram. Soc.*, Vol. 77, No. 3, pp. 843-46, ISSN 0002-7820

Nam, J. G., et al. (2001). Synthesis and sintering properties of nanosized In_2O_3-10 wt.-% SnO_2 powders. *Scripta Mater.*, Vol. 44, No. 8-9, pp. 2047-2050, ISSN 1359-6462

Nayak, J., et al. (2007). Yellowish-white photoluminescence from ZnO nanoparticles doped with Al and Li. *Superlattices and Microstructures*, Vol. 42, No. pp. 438–443, ISSN 0749-6036

Nichols, D. R. (1982). ITO films: adaptable to many applications. *Photonics Spectra*, Vol. No. pp. 57-60, ISSN 0731-1230

Odaka, H., et al. (2001). Electronic structure analyses of Sn-doped In_2O_3. *Jpn. J. Appl. Phys.*, Vol. 40, No. 5A, pp. 3231-3235, ISSN 0021-4922

Otsuka-Matsua-Yao, S., et al. (1997). Chemical hysteresis on the release and uptake of oxygen by SnO_2-doped In_2O_3 powders. *J. Electrochem. Soc.*, Vol. 144, No. 4, pp. 1488-1494, ISSN 0013-4651

Patzke, G. R., et al. (2000). Chemischer Transport fester Loesungen. 8. Chemischer Transport und Sauerstoffionenleitfaehigkeit von Mischkristallen im System In_2O_3/SnO_2. *Z. Anorg. Allg. Chem.*, Vol. 626, No. 11, pp. 2340-2346, ISSN 1521-3749

Piticescu, R. R., et al. (2006). Synthesis of Al-doped ZnO nanomaterials with controlled luminescence. *J. Eur. Ceram. Soc.*, Vol. 26, No. pp. 2979–2983, ISSN 0955-2219

Rahaman, M. N. (1995). *Ceramic processing and sintering*, Marcel Dekker, Inc., ISBN 0-8247-9573-3, New York

Schlott, M., et al. (2001), Method of preparing indium oxide/tin oxide target for cathodic sputtering, US 6187253, Leybold Materials GmbH, Patent

Schlott, M., et al. (1996). Nodule formation on indium-oxide tin-oxide sputtering targets, *Proceedings of 1996 SID International Symposium*, Santa Anna, CA, USA

Schlott, M., et al. (1995), Target für die Kathodenzerstäubung zur Herstellung transparenter, leitfähiger Schichten und Verfahren zu seiner Herstellung, DE 4407774, Patent, Leybold Materials GmbH

Selmi, M., et al. (2008). Studies on the properties of sputter-deposited Al-doped ZnO films. *Superlattices and Microstructures*, Vol. 44, No. pp. 268–275, ISSN 0749-6036

Shui, A., et al. (2009). Preparation and properties for aluminum-doped zinc oxide powders with the coprecipitation method. *J. Ceram. Soc. Japan*, Vol. 117, No. 5, pp. 703-705, ISSN 1882-1022

Shunk, F. A. (1969). *Constitution of binary alloys* (2 supplement), McGraw-Hill, Inc., ISBN 07-057315-8, Chicago

Son, J. W. & Kim, D. Y. (1998). Enhanced densification of In_2O_3 ceramics by presintering with low pressure (5 MPa). *J. Am. Ceram. Soc.*, Vol. 81, No. 9, pp. 2489-2492, ISSN 0002-7820

Sorescu, M., et al. (2004). Nanocrystalline rhombohedral In_2O_3 synthesized by hydrothermal and postannealing pathways. *J. Mater. Sci.*, Vol. 39, No. pp. 675-677, ISSN 1573-4838

Stanciu, L. A., et al. (2001). Effects of Heating Rate on Densification and Grain Growth during Field-Assisted Sintering of α-Al_2O_3 and $MoSi_2$ Powders. *Metall. Mater. Trans. A*, Vol. 32, No. pp. 2633-2638, ISSN 1073-5623

Stenger, B., et al. (1999), Verfahren zum Herstellen eines indium-Zinn-Oxid-Formkörpers, DE 19822570, Patent, W.C. Heraeus GmbH

Strachowski, T., et al. (2007). Morphology and luminescence properties of zinc oxide nanopowders doped with aluminum ions obtained by hydrothermal and vapor condensation methods. *J. Appl. Phys.*, Vol. 102, No. 7, pp. 073513 - 073513-9 ISSN 0021-8979

Straue, N., et al. (2009). Preparation and soft lithographic printing of nano-sized ITO-dispersions for the manufacture of electrodes for TFTs. *J. Mater. Sci.*, Vol. 44, No. 22, pp. 6011-6019, ISSN 1573-4838

Takeuchi, T., et al. (2002). Rapid preparation of indium tin oxide sputtering targets by spark plasma sintering. *J. Mater. Sci. Lett.*, Vol. 21, No. pp. 855-857, ISSN 0261-8028

Thu, T. V. & Maenosono, S. (2010). Synthesis of high-quality Al-doped ZnO nanoink *J. Appl. Phys.* , Vol. 107, No. 1, pp. 014308-014308-6, ISSN 0021-8979

Udawatte, C. P. & Yanagisawa, K. (2001). Fabrication of low-porosity indium tin oxide ceramics in air from hydrothermally prepared powders. *J. Am. Ceram. Soc.*, Vol. 84, No. 1, pp. 251-53, ISSN 0002-7820

Udawatte, C. P. & Yanagisawa, K. (2000). Hydrothermal preparation of highly sinterable tin doped indium oxie powders: the effect of the processing parameters, *Proceedings of Ceramic Processing Science VI*, Inuyama City, Japan,

Udawatte, C. P., et al. (2000). Sintering of additive free hydrothermally derived indium tin oxide powders in air. *J. Solid State Chem.*, Vol. 154, No. 2, pp. 444-450, ISSN 0022-4596

Vojnovich, T. & Bratton, R. J. (1975). Impurity effects on sintering and electrical resistivity of indium oxide. *Am. Ceram. Soc. Bull.*, Vol. 54, No. 2, pp. 216-217, ISSN 0002-7812

Warschkow, O., et al. (2003). Defect structures of tin-doped indium oxide. *J. Am. Ceram. Soc.*, Vol. 86, No. 10, pp. 1700-1706, ISSN 0002-7820

Weigert, M., et al. (1992), Target für die Kathodenzerstäubung und Verfahren zu dessen Herstellung, DE 4124471, Patent, Degussa AG

Werner, J., et al. (1996). Chemical transport of restricted solid solutions of In_2O_3 and SnO_2: experiments and thermodynamic process analysis. *J. Cryst. Growth*, Vol. 165, No. 3, pp. 258-267, ISSN 0022-0248

Wit, J. H. W. d. (1975). The high temperature behaviour of In_2O_3. *J. Solid State Chem.*, Vol. 13, No. pp. 192-200, ISSN 0022-4596

Yanagisawa, K., et al. (2000). Preparation and characterization of fine indium tin oxide
    powders by a hydrothermal treatment and postannealing method. *J. Mater. Res.*,
    Vol. 15, No. 6, pp. 1404-1408, ISSN 0884-2914

Yanai, Y. & Nakamura, A. (2003), Manufacturing method of ITO powder with tin dissolved
    in indium oxide, and manufacturing method of ITO target, US 2003/0039607,
    Patent,

Permissions

The contributors of this book come from diverse backgrounds, making this book a truly international effort. This book will bring forth new frontiers with its revolutionizing research information and detailed analysis of the nascent developments around the world.

We would like to thank Arunachalam Lakshmanan, for lending his expertise to make the book truly unique. He has played a crucial role in the development of this book. Without his invaluable contribution this book wouldn't have been possible. He has made vital efforts to compile up to date information on the varied aspects of this subject to make this book a valuable addition to the collection of many professionals and students.

This book was conceptualized with the vision of imparting up-to-date information and advanced data in this field. To ensure the same, a matchless editorial board was set up. Every individual on the board went through rigorous rounds of assessment to prove their worth. After which they invested a large part of their time researching and compiling the most relevant data for our readers. Conferences and sessions were held from time to time between the editorial board and the contributing authors to present the data in the most comprehensible form. The editorial team has worked tirelessly to provide valuable and valid information to help people across the globe.

Every chapter published in this book has been scrutinized by our experts. Their significance has been extensively debated. The topics covered herein carry significant findings which will fuel the growth of the discipline. They may even be implemented as practical applications or may be referred to as a beginning point for another development. Chapters in this book were first published by InTech; hereby published with permission under the Creative Commons Attribution License or equivalent.

The editorial board has been involved in producing this book since its inception. They have spent rigorous hours researching and exploring the diverse topics which have resulted in the successful publishing of this book. They have passed on their knowledge of decades through this book. To expedite this challenging task, the publisher supported the team at every step. A small team of assistant editors was also appointed to further simplify the editing procedure and attain best results for the readers.

Our editorial team has been hand-picked from every corner of the world. Their multi-ethnicity adds dynamic inputs to the discussions which result in innovative outcomes. These outcomes are then further discussed with the researchers and contributors who give their valuable feedback and opinion regarding the same. The feedback is then collaborated with the researches and they are edited in a comprehensive manner to aid the understanding of the subject.

Apart from the editorial board, the designing team has also invested a significant amount of their time in understanding the subject and creating the most relevant covers. They scrutinized every image to scout for the most suitable representation of the subject and create an appropriate cover for the book.

The publishing team has been involved in this book since its early stages. They were actively engaged in every process, be it collecting the data, connecting with the contributors or procuring relevant information. The team has been an ardent support to the editorial, designing and production team. Their endless efforts to recruit the best for this project, has resulted in the accomplishment of this book. They are a veteran in the field of academics and their pool of knowledge is as vast as their experience in printing. Their expertise and guidance has proved useful at every step. Their uncompromising quality standards have made this book an exceptional effort. Their encouragement from time to time has been an inspiration for everyone.

The publisher and the editorial board hope that this book will prove to be a valuable piece of knowledge for researchers, students, practitioners and scholars across the globe.

List of Contributors

Gislâine Bezerra Pinto Ferreira, José Ferreira da Silva Rubens Maribondo do Nascimento, Uílame Umbelino Gomes and Antonio Eduardo Martinelli
Federal University of Rio Grande do Norte, Brazil

Adriana Scoton Antonio Chinelatto, Ana Maria de Souza, Milena Kowalczuk Manosso and Adilson Luiz Chinelatto
Department of Materials Engineering - State University of Ponta Grossa, Brazil

Elíria Maria de Jesus Agnolon Pallone
Department of Basic Sciences – FZEA - São Paulo University, Brazil

Roberto Tomasi
Department of Materials Engineering - Federal University of São Carlos, Brazil

R.A. Vargas-Ortíz, F.J. Espinoza-Beltrán and J. Muñoz-Saldaña
Centro de Investigación y de Estudios Avanzados del IPN, Unidad Querétaro, Libramiento Norponiente No. 2000, Fracc. Real de Juriquilla, CP Querétaro, Qro., México

Zongqing Ma and Yongchang Liu
Tianjin Key Lab of Composite and Functional Materials, School of Materials Science & Engineering,
Tianjin University, Tianjin, P R China

Marta Suárez and Adolfo Fernández
ITMA Materials Research
Centro de Investigación en Nanomateriales y Nanotecnología (CINN). Consejo Superior de Investigaciones Científicas (CSIC) – Universidad de Oviedo (UO) – Principado de Asturias, Spain

Ramón Torrecillas and José L. Menéndez
Centro de Investigación en Nanomateriales y Nanotecnología (CINN). Consejo Superior de Investigaciones Científicas (CSIC) – Universidad de Oviedo (UO) – Principado de Asturias, Spain

Irene Barrios de Arenas
Instituto Universitario de Tecnología, "Dr Federico Rivero Palacio", Venezuela

Zarbout Kamel and Kallel Ali
Sfax University, LaMaCoP, BP 1171, Sfax 3000, Tunisie

Moya Gérard and Si Ahmed Abderrahmane
Aix-Marseille University, Im2np, UMR-CNRS 6242, Marseille, France

Damamme Gilles
Commissariat à l'Energie Atomique, DAM Ile-de-France, Bruyère-le-Châtel, France

V.P. Ilyina
Establishment the Karelian Centre of Science of the Russian, Academy of Sciences Institute of Geology of the Russian, Academy of Science, Russia

Guido Falk
Saarland University, Chair Powder Technology of Glass and Ceramics, Saarbruecken, Germany

Printed in the USA
CPSIA information can be obtained
at www.ICGtesting.com
JSHW011413221024
72173JS00004B/529